Cruis agement

Cruise Operations Management

Philip Gibson

AMSTERDAM • BOSTON • HEIDELBERG • LONDON
NEW YORK • OXFORD • PARIS • SAN DIEGO
SAN FRANCISCO • SINGAPORE • SYDNEY • TOKYO
Butterworth-Heinemann is an imprint of Elsevier

Butterworth–Heinemann is an imprint of Elsevier
30 Corporate Drive, Suite 400, Burlington, MA 01803, USA
Linacre House, Jordan Hill, Oxford OX2 8DP, UK

Library of Congress Cataloging-in-Publication

Gibson, Philip, 1955-
Cruise operations management / Philip Gibson.
p. cm.
Includes bibliographical reference (p.).
ISBN 0-7506-7835-6 (alk. paper)
1. Ocean travel–Management. 2. Cruise lines–Management. I. Title.

G550.G53 2006
387.5'42068–dc22 2005058172

British Library Cataloguing-in-Publication Data
A catalogue record for this book is available from the British Library.

ISBN 13: 978-0-7506-7835-3
ISBN 10: 0-7506-7835-6

For information on all Butterworth–Heinemann publications
visit our Web site at www.books.elsevier.com

Printed in the United States of America

06 07 08 09 10 10 9 8 7 6 5 4 3 2 1

Contents

Preface

A generation ago, I was employed by P&O Cruises in the purser's department on the "old" Oriana. I was blessed with youth, naivety, and enthusiasm and was eager to embrace the world. Since then, my career in the tourism and hospitality industries, and subsequently in further and higher education, has been rewarding and fulfilling, but I still cast my memory back to those earlier times to reflect on the excitement, the quality of life, and the fun that was engendered while working on board a luxury cruise ship.

In those days, cruising was a traditional pastime and yet, even then, there was change on the horizon. Customers were notably different in demographic terms when comparing sailing from Southampton and sailing from Sydney. The United States was emerging as the largest growing market, the trend being fanned by a popular television show of the time called *The Love Boat*. Ominously, cruise ships were subject to threats of terrorism, adding a certain gravity to emergency drills.

The constant, however, was the accessibility to awe-inspiring sights: drifting past jungles and wildlife-rich terrain while traversing the Panama Canal, sailing into Sydney harbor alongside the famous Opera House, anchoring off picturesque silver-sanded Caribbean islands in warm azure seas. This was a unique job and the levels of customer satisfaction were high.

While some young people's aspiration to work at sea has appeared to be rather muted, the rewards of working on a cruise ship continue to provide a peculiar attraction. The strength of growth in the sector is undiminished, and this type of career provides a powerful option to satisfy cravings for travel while stimulating the development of new dreams.

This book reflects an amalgam of those elements that are relevant to someone seeking to work on a modern cruise ship. My short career at sea has most certainly been an important factor in preparing to write this book, but I believe the support from professionals in the industry has been even more significant and I am very grateful to all who helped. As a postscript, I was recently traveling through China for my university, speaking with representatives from educational institutions about student progression to study in the UK. On the final leg of my journey I visited a town called Dalian, which is situated in northeast China. I was astonished to see a rather old but patently impressive (in a faded glory type of way) cruise ship moored to a centrally located and wide-open shoreside area. It was "my" *Oriana*. Sadly, I learned that she (ships are generally referred to as being female) was severely damaged in a storm in June 2004 and her prognosis, despite efforts to repair the damage, is not favorable.

Acknowledgments

In order to write this book, I have interviewed in excess of 100 cruise industry professionals including passenger service directors, executive pursers, staff pursers, training managers, buffet stewards, executive head chefs, corporate chefs, accommodation managers, head waiters, junior assistant pursers, assistant managers, shop managers, photography managers, cruise directors and laundry masters.

I am indebted to many people, including those I have interviewed, for their time and support. In particular I would like to thank Brian Johnson of Princess Cruises, Rick Godwin and Celia Walters of P&O Cruises, and the officers and crew of the *Star Princess*, the *Golden Princess*, and the *Aurora*. I am also very grateful to Kevin Spencer of Cunard, Bob Harrison of Destination Southwest, and my colleagues at the University of Plymouth for their support.

Finally, but most importantly, I would like to thank my daughters Hazel and Rachel for acting as pseudo-student reviewers in reading the draft chapters, and my wife Carol, a former assistant purser, for her seal of approval.

Introduction

This textbook is directed at those who wish to study hotel operations management on cruise ships. There is a dearth of literature focusing on contextualized operational management on board cruise ships. In part this is because the industry has undergone a quiet renaissance over the last two decades, steadily increasing in scale and scope without the fanfare and academic attention that accompanies land-based tourist activities. The unique characteristics of the industry are such that a customized book is essential for those seeking to study the industry, whether for academic or vocational reasons. It is suggested that this text will also support students involved in postgraduate degree work.

This book is designed to provide a comprehensive overview of hotel services for the cruise industry. Using examples from cruise operations, the reader will be in a position to consider those aspects of this industry that are unique and to draw on the information provided to create a clear understanding about managing operations on board.

As well as presenting a range of contextualized facts, the book includes a number of case studies that encourage the reader to examine the often complex circumstances that surround problems or events associated with cruise operations. The case studies are reflections of reality on board cruise ships, constructed by using a combination of interviews and observations. This research was undertaken with the support of a number of cruise companies.

These cases include an example of food and drink operations on board a large cruise ship that caters predominantly to US customers on vacation. Other case studies describe innovation in training as introduced by Cunard, the management and operation of administration on board "mega" cruise ships operated by Princess Cruises and P&O Cruises, accommodation management, organizing the servicing of staterooms, operating shore excursions, and providing special services for passengers. In addition, a series of case reports highlight cultural factors relating to crew members on board. Case studies are followed by questions, which are intended to illuminate issues and stimulate discussion.

The chapters are organized to support systematic development of knowledge. The first chapter creates a sound introduction to the industry in its social, technological, and economic setting. The second chapter considers the current infrastructure for selling cruises. The third focuses on maritime issues, describing the scope and scale of the industry in terms of the world fleet and reflecting on critical factors such as the framework for international law and environmental concerns. The fourth chapter examines cruise geography in order to identify cruise sectors and major destinations. This knowledge is then further developed in chapter five, which considers itinerary planning. Practices are explained that itinerary planners can use when considering cruise content such as ports of call shore excursions. These examples help to form a full picture in relation to this part of the cruise product.

Chapter six relates to management and operational structures on board a cruise ship. Chapter seven examines customer or guest services in depth to provide commentary on the range of services that are provided on many cruise ships. Chapters eight and nine consider subjects that are central for hotel services: the management of food and drink operations and the management of onboard facilities. Chapter ten reflects on health, safety, and security, providing a contemporary setting for these challenging topics. Chapter eleven debates training issues in general, with special regard to the notion of situated learning. Chapter twelve provides an integrated conclusion to management of cruise operations with an appraisal of administrative functions on board and oversight of how the dynamics of a cruise ship pull together to establish successful operations.

While the content of each chapter is intended to be cumulatively informative and in many ways concerned with theory building, students, lecturers, and professionals in the cruise industry can derive benefit from adopting techniques that are described and suggested as being good practice. In this way the book can be relevant holistically for those intending to learn about managing cruise operations and of value to those who wish to dip in and out of chapters to make use of specific material within a tourism or hospitality context.

1

Contemporary Cruise Operations

Learning Objectives

By the end of the chapter the reader should be able to:

- Define the elements of cruising
- Be aware of the history of cruising
- Critically reflect on the image of cruising and consider different types of cruises
- Understand the scale and scope of the cruise market

The cruise industry has grown and continues to grow enormously in scale. It is frequently regarded as a small but significant sector of the tourism industry (Page, 2002), but this description is insufficient in recognizing inherent qualities and attributes that support the claim that cruising is an industry in its own right. In many respects it is helpful to consider some evidence regarding this claim within this introduction, but readers will be able to make a more informed judgment having read the whole book.

According to Ward (2001), the cruise industry has a $15 billion turnover. It employs over 100,000 shipboard officers and crew as well as approximately 15,000 employees ashore. Indirectly, the industry provides employment for food suppliers, engineering services, manufacturers, port agents and authorities, transport companies, tourist companies, hotels, destination companies, and car-hire and employment agencies. Ebersold (2004) draws attention to the growth of the industry in marketing terms, which has seen systematic and sustained expansion over 7 years, with approximately 64 million bed-days sold in 2002 compared with 35 million bed-days sold in 1997. Construction of new vessels for the industry continues to be strong, and at the beginning of 2004 there were some 19 new cruise ships under construction, catering to a total of approximately 48,000 passengers (Bond, 2004). The Cruise Lines International Association (CLIA, 2005b), an organization that represents 19 of the world's major cruise companies, announced that 10.5 million passengers chose to cruise in 2004. The changing demographic profile of cruising, in terms of the market segment, social status, and age, is significant (Douglas and Douglas, 2004). One example of these changes is the estimate made by Carnival Cruises that they would carry some 500,000 children as passengers in 2004: an increase of 400% over 10 years (Carnival, 2005).

In 2005, 11.1 million passengers are expected to take a cruise with one or more of the CLIA's member companies (Anon, 2005a), a projected 4.6% increase over 2004 figures. In 2004, 12 new ships were introduced, accounting for a 6.9% increase in capacity. The CLIA predicts that cruise fares will increase, that the age of those cruising will continue to reflect a multigenerational mix, and that all segments of the cruise market will become even more focused on providing unique products and services (Anon, 2005a).

The Elements of Cruising

It could be argued that our planet Earth is, in one significant sense, misnamed. This is because 71% of the surface is covered by water (Lutgens, 1992). Air travel has been cited as a major influence on changing leisure activities, yet even a novice can recognize the opportunities for sea and water based vacations using ships as floating resorts. According to Day and McRae (2001), a cruise ship provides easy access to some of the world's most popular destinations, and this simple statement holds the key to the current successes that the industry enjoys. This can be exemplified by examining Table 1.1 and completing the task that is described.

For many tourists, the cruise experience embodies a series of powerful motivators: it is often perceived to be safe, social, customer friendly, and service oriented (Cartwright & Baird, 1999). The ship provides a mobile, consistent, and easily accessible location to act as a home away from home while the tourist samples the port of call. The tourist adapts to shipboard life and learns to relax into a vacation routine (Gibson, 2003): a routine that can be interspersed with a choreographed range of ship or land activities.

As travel expert Douglas Ward says, "Over 10 million people can't be wrong (that's how many people took a cruise last year)! Cruising is popular today because it takes you away from the pressures and strains of contemporary life by offering an escape from reality. Cruise ships are really self-contained resorts, without the crime, which can take you to several destinations in the space of just a few days" (Ward, 2001).

However, the notion of "cruising" also generates negative perceptions (Table 1.2). Dickinson and Vladimir (1997) conducted interviews with people who either had not considered or did not want to go on a cruise. They revealed five specific factors that demotivated the potential tourist: Ward (2001) counters this list by highlighting emerging patterns. Cruising is presented as being both cost effective and high in value. The range of cruise types has expanded to include opportunities for all sorts of people. In this way, cruising can be both socially inclusive and exclusive: families can be catered to as a specialty market, as can single tourists, conference delegates, older travelers, active tourists, groups, etc.—the list is endless. Ward recognizes that this type of vacation is appealing to older customers but also notes that the average age of first time cruisers is now well under 40.

Table 1.1: **Tourists' favorite overseas city**

Using the Observer/Guardian's travel Awards list tourists' "favourite overseas city" (Anon, 2003a), identify those that can, in theory, be visited by a cruise ship.

1. Sydney
2. Melbourne
3. Tokyo
4. Cape Town
5. Vancouver
6. Rio De Janeiro
7. Chicago
8. Dubai
9. Oslo
10. Orlando
11. New York
12. Perth
13. Venice
14. Hong Kong
15. Berlin
16. Prague
17. Bologna
18. Singapore
19. Barcelona
20. Rome
21. Kuala Lumpur
22. San Francisco
23. Las Vegas
24. Auckland
25. Vienna
26. Stockholm
27. Beijing
28. Budapest
29. Toronto
30. Bruges
31. Florence
32. Helsinki
33. Lisbon
34. Istanbul
35. Reykjavik
36. Munich
37. Verona
38. Seville
39. Copenhagen
40. Sienna
41. Havana
42. Bangkok
43. Salzburg
44. Bilbao
45. Madrid
46. Marrakech
47. Granada
48. Washington
49. Boston
50. Funchal

Table 1.2: **Factors that demotivate potential cruisers (adapted from Dickinson and Vladimir (1997))**

Factor	*Reason*
Cost:	cruising is perceived to be expensive
Exclusivity:	cruising thought to be a domain for the wealthy and elitist in terms of social groupings
Family prohibitive:	cruising not felt to be for people with children but rather oriented toward older couples
Claustrophobic:	the ship is thought of as a constraint and quiet space a premium
Seasickness:	concerns about coping with seasickness influence decision-making

A History of Cruising

Much can be gained in charting the history of cruising to identify not only where and how the concept of cruising arose, but also to try and predict where it is going. Table 1.3 (sources: Dickinson and Vladimir [1997], Knotes [2003], Michaelides [2003], Dawson [2000], Cartwright and Baird [1999], Day and McRae [2001], Showker and Sehlinger [2002] and Ward [2001]) is not intended to be inclusive but rather to chart significant moments over the last 200 years.

Much is said about the size of contemporary super-cruisers. The example in Table 1.3 of the introduction of "Eagle" class cruise ships, leading up to the launch of the *Queen Mary 2* (*QM2*) is a case in point. The ship as a destination with sophisticated on-board facilities and a much-enhanced product is linked to economies of scale achieved through the construction of larger vessels (Kontes, 2003). This aspect of cruising has captured the public's attention, and the implications are important in terms of political, economic, social, technological, legal and environmental issues. These aspects will be examined later in this book. However, even in basic terms, the sizes of cruise ships provide an interesting comparison (Figure 1.1).

Currently, the largest vessels can carry around 4,000 customers and the smallest fewer than 100 customers. The Cunard Line's *Queen Elizabeth 2* measures 70,327 GRT (Gross Registered Tonnage) and the *Queen Mary 2*, which started cruising in January 2004, is 150,000 GRT. Princess Cruises' *Grand Princess*, weighs in at 108,806 GRT, while Hebridean Island Cruises' *Hebridean Princess* is 2,112 GRT. Royal Caribbean's *Freedom of the Sea* (158,000 GRT) is due to embark on her inaugural cruise in April 2006. Scale varies depending on purpose. Large vessels, such as the Eagle-class ships that were discussed previously, accommodate larger numbers and can provide opportunities for greater diversity on board. Smaller vessels can be more intimate and provide access to ports that larger ships cannot visit because of the depth of the ship's keel, the length of the vessel, or the constraints of maneuverability at the destination. Certain ratios (crew to customers, customer space per customer, and size of cabin to public areas) all play a part depending on the type of cruise tourist. The passenger/space ratio is calculated by dividing GRT by the maximum number of passengers to provide a number that defines cubic space per passenger. Currently the *QM2* has one of the highest ratios of space to passengers at just over 57 (150,000 GRT divided by 2,620 passengers). At the other end of the scale, budget vessels might be as low as 28. Ratios of crew to passengers tend to reflect a 2:1 ratio for premium lines and 1.5:1 for luxury vessels.

The Image of Cruising

The industry is diverse, and it appears that this is indicative of the future direction for cruise developments. The following case studies present four contrasting cruise experiences. The cases are preceded by Table 1.4 that provides an easy comparison between basic features. The flag of registration is important because it refers to the legal status of the ship (see Chapter 3 for more information).

Table 1.3: **A history of cruising (various sources)**

Year	*Event*
1801	The tug '*Charlotte Dundas*' goes into service and becomes the first practical steam-driven vessel.
1818	Black Ball Line introduces the '*Savannah*' 424 GRT or Gross Registered Tonnage (GRT — see Figure 1.1 for an explanation of this term), carrying 8 customers and this ship becomes the first to cross the Atlantic from New York to Liverpool. The journey took 28 days.
1835	First advertised cruise around Shetland and Orkneys. This was a cruise that never occurred and it wasn't until 1886 that the North of Scotland and Orkney and Shetland Shipping Company operated short break cruises.
1837	Peninsular Steam Navigation company founded (later to become the Peninsular and Oriental, Steam and Navigation Company and the familiar name of P&O Cruises).
1840	Sam Cunard establishes the first transatlantic steamship.
1843	Isambard Kingdom Brunel's ship the '*Great Britain*', 3270 GRT, is launched. This ship is the first iron hulled, propeller driven customer vessel.
1844	P&O cruises from London to Vigo, Lisbon, Malta, Istanbul, and Alexandria aboard the 'SS *Iberia*'.
1858	Customers pay to join the '*Ceylon*', a P&O vessel, for what is considered the first cruise.
1867	Author Mark Twain features a P&O voyage from London to the Black Sea in his novel 'The innocents abroad'.
1881	The '*Ceylon*' is refitted to become a purpose built customer ship.
1910	White Star introduces the '*Olympic*' 46,329 GRT and, the year after, the *Titanic*, which sank having collided with an iceberg on 12 April 1912.
1911	'*Victoria Louise*' becomes the first vessel to be built exclusively for cruising.
1912	Cunard introduces the '*Laconia*' and '*Franconia*' as custom built cruise and line voyagers.
1920–33	In the USA during prohibition, 'booze cruises' from US ports allows customers to drink and gamble while visiting ports such as Cuba, Bermuda and the Bahamas.
1922	Cunard's '*Laconia*' sails on a world cruise. This ship was relatively small, 20,000 GRT and 2,000 customers in three class accommodation.
1929	P&O's '*Viceroy of India*' introduced. It was the most impressive ship of the time featuring the first use of turbo-electric-power and the first onboard swimming pool. It was a dual-purpose liner (UK to India) and luxury cruiser.
1930s	Union Castle offers holiday tours to South Africa at highly competitive rates of £30 third class, £60 second class and £90 first class.
1934	The luxury cruise liner 'RMS *Queen Mary*' is launched. With 1,174 officers and crew and 2,000 customers, the ratio is less than 2:1.
1934	United States lines builds the 'SS *America*', an oil fired liner capable of speeds up to 25 knots. This vessel is commissioned as a troop carrier in 1941.
1938	'SS *Normandie*' 83,000 GRT undertakes a 21 day cruise: New York-Rio de Janeiro-New York. Cost per customer from $395 to $8,600.
1939	World war two is declared. Cruise ships such as '*Queen Mary*' and '*Queen Elizabeth*' are converted as troop carriers.
1958	First transatlantic commercial jet–aircraft crossing leading to the demise of the liner market and the down-turn of business for many cruise companies.
1966	Cruise industry recovers — mainly centered on the UK.
1970s	New cruise companies established — 1% of holidaymakers take cruise holidays. Cruise companies work closely with airlines to develop combined products — fly-cruise.
1986	'*Windstar*' a vessel with computerized sails is introduced marrying the romance of sail with modern comforts.
1990s	Consolidation and globalization: leading to mergers and acquisitions.
1999	'Eagle' type vessels such as '*Voyager of the Sea*' and '*Grand Princess*' are introduced bringing higher levels of sophistication, economy of scale and the concept of the vessel as a destination.
2000s	Segmentation and Lifestyle cruising. Sustained growth for the North American market (8% annually) from 1980 to 2000.
2000	Royal Caribbean International's (RCI) '*Explorer of the Sea*' is introduced (137,308 GRT)
2002	There are an estimated 700 million tourists worldwide of whom 10.3 million are cruise tourists. 2.4% of US population, 1.3% of UK population and less than 1% of Europe's population cruise annually
2003	Cunard's '*Queen Mary 2*' launched (150,000 GRT).
2003	Carnival Corporation becomes the largest cruise operator when they merged with P&O Princess Cruises.

How do you measure the size of a ship?

Ships can be described by referring to capacity, dimensions or tonnage.

Capacity
A cruise ship's capacity is expressed in terms of the total numbers of officers, crew and customers. Cruise companies frequently plan using lower berth capacity (referring to the number of beds in a cabin), implying that capacity for some ships could be increased if capacity included upper berths (some cabins can have bunk-beds or two tier bedding arrangements).

Dimensions
The length is measured from the bows or forward end (fore) to the stern or after end (aft)—fore and aft are commonly used terms.
The beam is the width at the widest point (amidships).
The draft or draught of a ship measures the depth of a ship as the vessel sits in the water.

Tonnage
Ships tend to be described and compared in terms of gross registered tonnage (GRT). According to Branch (1996). GRT is calculated by dividing the volume in cubic feet of a vessel's closed-in spaces by 100. A vessel ton is100ft^3. Tonnage is frequently made use of by port authorities when calculating charges when a ship requires a pilot and for harbor fees. The word 'tonnage' is derived from 'tun' a medieval term meaning barrel.

Speed
Speed is measured in knots. 1 Knot equates to 1 nautical mile per hour. A nautical mile is the equivalent to 1,852 meters or 1.15 land miles.

Figure 1.1: Ship measurements

Windstar

The "Windstar" concept was introduced in 1986 (Figure 1.2). The aim was to combine the luxury of a vacation with the freedom of sailing to create a unique product. The company describes the concept as being "the ultimate getaway for tired executives, adventure seekers and self-described escapists" (Windstar 2003). The vessel '*Windstar*' shares her name with the company. She is one of three of the company's ships designed to cruise a range of destinations including the Lesser Antilles, Virgin Islands, Florida Keys and Bahamas in the Caribbean, and Western and Eastern Europe, the Mediterranean, the Greek and Turkish Isles, the Baltic and Northern Europe. Customers are described as having an average age of 51 for past guests and 50 for new guests, with an average annual income of $120,000 and over. Windstar customers are said by the company to be "active and adventurous, sophisticated travelers ranging from their 20s to 80s" (Windstar, 2003). Typically, Windstar customers may be business owners, executives, retirees, honeymooners, stockbrokers, lawyers, engineers, entrepreneurs, artists, authors, researchers, doctors, or educators.

According to Ward (2001), the ship's age creates both a luxury ambience through its use of materials and overall decor and, in the cabins, a more traditional appearance. The overall cruising experience is reported on positively. The experience is said to be relaxing, informal, and stress free. A range of activities, including wind surfing, skiing, scuba, and snorkeling, is available. The ship boasts a casino, a library, a small swimming pool (described by Ward as a "dip pool"), a shop, an infirmary, a piano bar, and a "chic and elegant" restaurant (called "the Restaurant"). Cabins are fitted out with television, videocassette and CD players, personal safe, refrigerator, mini-bar, and international direct-dial phones, as well as private bathrooms with shower toiletries, hair dryer, vanity, and plush robes.

The crew is described as being "international." The Captain and nautical staff are European. Hotel senior staff members are American and European. Stewards and service personnel are Filipino and

Table 1.4: **Comparison chart**

Vessel	*Windstar*	*World of ResidenSea*	*Grand Princess*	*Ocean Village*	*Queen Mary 2*
Operating company	Windstar Cruises	ResidenSea Ltd	Carnival	Carnival	Carnival
Built	1986	2002	1998	1989 (formerly Arcadia)	2003
GRT	5,350	43,524	109,000	63,500	150,000
Draft	4.1 meters	6.7 meters	8 meters	8.2 meters	10.09
Length	134 meters	196.35 meters	292 meters	247 meters	348 meters
Beam	15.8 meters	29.8 meters	36 meters	32 meters	45.4 meters
Speed	14 knots	18.5 knots	24 knots	21.5 knots	26 knots
Method of propulsion	Diesel electric (3) and sail (6)	Diesel electric	Diesel electric	Diesel electric	Gas turbine Diesel electric
Customer space ratio	36 cubic feet per pax	66 cubic feet per pax	42 cubic foot per pax	37.5 cubic feet per pax	56.25 cubic feet per pax
Customer cabins	74	110 apartments & 88 studios	1,300	801	1,310
Number customers	148 (based on double occupancy)	656 (average is expected to be 320)	2,600	1,692	2,620
Number crew	90	320	1,100	514	1,253
Marketing slogan	'180 degrees from ordinary'	'When you live to travel'	'Personal choice cruising'	'For people who don't do cruises'	'The world's largest, longest, tallest, grandest ocean liner ever'
Flag—country of registration	Bahamas	Bahamas	Liberia	United Kingdom	United Kingdom

Indonesian. In total, the design is intended to create an environment that is intimate, contemporary, and luxurious. The company emphasizes that the size of the ship, coupled with the facilities and service, easily achieves this aim.

The World

The World of ResidenSea (40,000 GRT) is a novel concept (Showker and Sehlinger, 2002). This vessel was built by the Fosen yard in Norway to continuously circumnavigate the world and to provide "the world's first ship to be designed as an ocean-going residence for full time occupancy . . . . The '*World*' provides spacious residences—fully furnished and equipped—and guest suites for family, friends, business associates, or personal staff" (Synnove Bye, 2003).

This ambitious project has dual occupancy options. The ship has 110 private apartments, each with a fully equipped kitchen (galley) that can be either privately owned or rented. In addition, the '*World*' has 88 guest suites that can be booked by the general public. Guests are expected to be 40 percent from the United States, 40 percent European, and 20 percent from the rest of the world. The target market for ownership is homeowners with two or three residences, average age 55, who are wealthy "self made" entrepreneurs. A typical profile would describe such a person as active, with a love of the sea or sailing and a desire to guard his or her privacy (ResidenSea, 2003).

The facilities include four distinctive restaurants, a night club, a casino, theatre, an art gallery, a spa and fitness center, two pools, a full-size tennis court, a golf center (including a real grass putting green), and a retractable marina. The ship also possesses three emergency hospital wards (Figure 1.3).

Figure 1.2: Windstar

Figure 1.3: The World of ResidenSea

The itinerary in a typical year includes 140 ports in 40 countries. The ship targets prestigious events, including sporting occasions such as the British Open, the Grand Prix in Monaco, and the Cannes Film Festival. The staff on board is international. The cost of purchasing an apartment begins at US $2,255,000. Vacations can be booked for as long as an individual wants, starting with a minimum of 3 days.

According to one of the company's own publications:

> The World of ResidenSea, as the first mixed-use resort ship continuously navigating the globe, set a new industry standard when it scored a perfect 100 points on its very first United States Public Health (USPH) inspection and received an excellent rating on its Certificate of Compliance from the United States Coast Guard. These reports echo international acclaim from Bahamian Authorities and Det Norske Veritas for the ship's health, safety, operational and construction standards, including comments that The World is "well-maintained and operated in a very professional manner".
>
> Contributing to the success of The World's recent ratings is its unique Scandinavian wastewater cleaning system, whereby wastes are filtered by a flotation system. Solid wastes are dried and incinerated, and the ash is properly disposed of on land. The remaining liquid waste goes through an ultraviolet filtration process, and the resulting water is as pure as technical water. The World also burns marine diesel, a departure from heavy fuels traditionally used in ships of its size and enables The World to enter more of the world's most fascinating ports (ResidenSea, 2003).

Grand Princess

The Grand Princess is part of a modern fleet of ships operated by Princess Cruises. Princess Cruises entered the market in 1965 with a single ship cruising to Mexico. Today, its fleet carries more than 800,000 customers each year. Ships include *Coral Princess* (2003), *Dawn Princess* (1997), *Island Princess* (2003), *Pacific Princess* (1999), *Regal Princess* (1991), *Royal Princess* (1984), *Golden Princess* (2001), *Grand Princess* (1998), *Star Princess* (2002), *Sun Princess* (1995), and *Tahitian Princess* (1999). Four additional new ships will join the Princess fleet by 2006 (Princess Cruises, 2003).

In 1977, the Pacific Princess was cast in a starring role on a new television show called *The Love Boat*. According to Princess Cruises, the weekly series, which introduced millions of viewers to the still-new concept of a sea-going vacation, was an instant hit, and both the company name and its "sea witch" logo (Figure 1.4) have remained synonymous with cruising ever since.

The Grand Princess was regarded in 2003 by Princess Cruises as being the "flagship of Personal Choice Cruising." This concept seeks to empower customers so that they could create their own selection of activities from a broad choice of facilities, amenities, and services. In this way the vacation could then be perceived as being personally customized. Keynote provision centers on multiple dining options, flexible and varied entertainment selections, and a full complement of on-board activities ranging from shuffleboard to scuba certification. Furthermore, the experience is presented using marketing slogans, such as "affordable luxury" and "big ship choice with small ship feel," to focus on the unique orientation of the vessel within the marketplace.

The ship is one of the largest in the world (although not the largest). Its great size enables the operator to include diversity and provide choice. Seven hundred and ten cabins (80 percent of all outside staterooms) have a balcony. Butler service is provided in suites and minisuites. The ship's facilities include a chapel, a virtual reality center, a casino, three dining rooms and three show lounges, five swimming pools (one is a swim-against-the-current lap pool), a children's center and a teenager center, 28 wheelchair accessible cabins, a sports bar, an art gallery, a nine-hole putting course and golf simulator, a sports court and jogging track, and a wide variety of bars and lounges, including a wine and caviar bar and the "skywalkers nightclub" and observation lounge suspended 150 feet above the water. As Showker and Sehlinger (2002) comment, size matters for this type of ship, with its 2,600 customers and 1,100 crew. Not only does the scale of the vessel support the scale of activities and facilities on offer, the ship travels well, providing a smooth ride.

Figure 1.4: Grand Princess

The Grand Princess offers voyages on a range of cruises including circular routes around the Caribbean. Smaller resorts are accessed by ship's tenders (the small launches that are carried by cruise vessels for both practical and safety reasons). Princess Cruises has ownership of a Caribbean island, "Princess Cays," which is accessed by tender. The ship anchors offshore and ferries passengers from ship to shore using ships' tenders. Princess Cruises is part of Carnival Corporation, one of the largest vacation companies in the world (Ward, 2001).

Ocean Village

Ocean Village is a relatively new brand introduced in 2003 by P&O Princess Cruises. Their target market is "middle youth," which is explained by Nick Lighton, managing director of Ocean Village, who states that "Ocean Village holidaymakers are much younger than traditional cruisers—our average booking age is 40—and our research shows they are looking for activity as well as relaxation."

Nick Lighton explains the thinking behind Ocean Village: "We researched exactly what 'middle-youth'—that is, 35–55 year olds with active modern lifestyles—were after in a holiday, and feedback showed they seek stimulating experiences as well as the chance to unwind. So Ocean Village cruises offer 'chill and challenge' holidays—not only can people take it easy in the usual sense but they can also return home having enjoyed a wealth of exciting experiences, be that mountain biking across

Barbadian beaches, exploring Antigua off the beaten track by jeep, or hiking through Dominica's rain-forests. . . . Ocean Village has ripped up the cruising rule book, letting customers eat when they want and with whomever they want, giving formal dress codes their marching orders and throwing traditional entertainment overboard" (Ocean Village, 2003).

Ocean Village is based in Palma, Majorca, for summer cruising in the Mediterranean and in Bridgetown, Barbados, for winter cruising in the Caribbean. From each port the ship embarks on two alternating itineraries that are designed to appeal to a target group. From Palma, the itineraries are called "Palaces and Paella" or "Piazzas and Pasta" and from Barbados the itineraries are referred to as "Sugar and Spice" or "Calypso and Coconuts." These holidays are designed to appeal to the target demographic. These itineraries are based around 7-day cruises to include one full day at sea and 6 days at a various resorts or ports.

The company emphasizes the notion of "cruises for people who don't do cruises" (Ocean Village, 2003), by presenting a ship with an informal style: no Captain's cocktail party, no dress code, and no need to eat at specific times. Leisure and entertainment activities are given a contemporary twist, with opportunities for "guests" to be as active or as inactive as they wish when onboard or ashore. Facilities onboard include a gym, two pools, four hot tubs, a spa, a jogging track, mountain bikes (to take ashore), a nightclub, a cinema, a circus and cabaret, a celebrity chef restaurant (extra charge), a kids' center, and eight bars.

In 2003 the company offered seven-night Mediterranean cruises from £499, while seven-night Caribbean cruises starting at £649. Customers paid extra for superior cabins, drinks, dining in the bistro or the deck cafe, shore activities, and some optional onboard activities such as Internet usage. The crew of 514, which comprises international hotel service staff and British officers and managers, provides a three-customer-to-crew member ratio. The success of this venture is underlined by the introduction of a second ship (Figure 1.5).

Figure 1.5: Ocean Village (courtesy of Graham Busby)

Queen Mary 2

The *QM2* entered service as a Cunard liner in January 2003 (Figure 1.6). She represented the epitome of scale and grandeur at sea by being the world's largest, longest, tallest, and grandest ocean liner. The ship's voluminous public areas, grand ballroom, staircases, and foyer areas and 360-degree promenade deck add credence to the claims for quality, luxury, image, and space. Three-quarters of the ship's staterooms have balconies. Apart from staterooms, the options for accommodation are wide, with alternatives including suites, junior suites, royal suites, penthouses, and duplexes. The two-story duplexes even have their own private exercise equipment. Accommodation on the *QM2* is linked to the type of dining experience. Guests who book a duplex, suite, or similarly graded accommodation have access to the exclusive Queens or Princess Grills, while guests who reserve staterooms may dine in the Britannia restaurant. The *QM2* restaurants, reflecting the Cunard attention to excellence, are classified as being of 5 star quality by the 2005 Berlitz Guide to cruising.

The ship has 14 decks containing sports facilities, shops, bars, lounges, five pools, and no fewer than 10 restaurants. There is a unique spa club, a casino, a planetarium, a bookstore, and a college-at-sea. As befits a ship of this scale, the options are very wide, providing cultural and artistic diversions during the day and theater productions, dancing, entertainment, a night club, a casino, and even karaoke by night. The ship is destined to take over from the *Queen Elizabeth 2* (*QE2*) to provide a luxury "line voyage" between New York and Southampton and to act as the flagship for Cunard's fleet. Scale may well play a significant part for the *QM2* in terms of the grandeur and opulence inferred by the ship's size and awe-inspiring statistics, but the other part of this equation concerns the ship's ability to access ports. According to Cruisemates.com (2005), about 50% of ports are likely to be "boat ports," where the ship will operate tenders to transport passengers from ship to shore.

The Cruise Market

The cruise industry in the twenty-first century is characterized by diversity and positive growth (Dingle, 2003). Traditional cruises exist, and indeed this market is strong (Michaelides, 2003), yet growth in customer demand is predicted to continue in less traditional markets.

Figure 1.6: The QM2

Table 1.5: **Segmentation in the cruise industry (Bjornsen, 2003)**

Segment	*Budget*	*Contemporary*	*Premium*	*Niche*	*Luxury*
Share	5%	59%	30%	4%	2%
Cruise duration	Varies	3–7 days	7–14 days	7 days and upwards	7 days and upwards
Ships	Older, smaller	New, large and mega	New, medium, and large	Small	Small and medium
Cruise lines	My Travel, Thomson, Royal Olympia	Carnival, Royal Caribbean, NCL, Princess, Costa, Royal Olympia, Ocean Village, Aida, Island Cruise, Arosa	Celebrity, Holland America, Cunard	A&K, Swan Hellenic, Star Clippers, Clipper, Lindblad Explorer, Orient Cruise Lines	Crystal, Silversea, Seabourn, Radisson 7 Seas, Seadream Yacht, Cunard, Windstar, Hapag Lloyd
Itinerary	Caribbean, Mediterranean, Baltic	Caribbean, Mediterranean	Caribbean, Mediterranean, Alaska	Worldwide, Antarctica, Greenland, Asia	Worldwide
Average cost per day (USD) per pax	80–125	100–50	150–300	200–900	300–2,000
Less Expensive ←					→ More Expensive

The implications of this analysis of the cruise market can appear contradictory. On one hand the numbers of traditional cruise customers appear to be both stable and increasing, yet there is clear evidence to suggest that real growth is likely to occur in emerging markets such as that described in the case study on the Ocean Village. In reality, cruise companies are continuing the trend that is presented in Table 1.5—that is, companies adapt to emerging situations and seek new opportunities to take advantage of trading conditions.

Acquisitions and Mergers

Cruising is undergoing change. The future is predicted to be good, and yet the last decade has been punctuated by a series of acquisitions and mergers (Bjornsen, 2003) that appear to indicate a market subject to constant upheaval and change. Table 1.6 lists some of the most significant examples.

In this environment it is tempting to forecast more of the same for the medium term but that would ignore salient issues that have emerged. First, the largest companies, the Carnival Corporation, Royal Caribbean Cruise Line, and Star Cruises/NCL, possess among them considerable purchasing power. Second, new ships create a wave effect, releasing into the market older vessels that can then be used by emerging or newly formed cruise companies. Third, cruise companies have learned to be wary of market conditions and unforeseen events. Public opinion of globalization, the incidence of terrorism, and worldwide threats caused by health problems such as severe acute respiratory syndrome (SARS), which emerged in China in 2002, can influence customers' decision making.

Recent industry commentaries demonstrate the resilience of the cruise industry in the face of a plethora of world events. As an example, reports on the top 10 cruise lines in this market posted a 9 percent increase in customer numbers for the first half of 2003 compared to the first half of 2002 for North American cruises, using data that was compiled by the US Maritime Administration (Anon, 2003). Customer figures for the second quarter of 2003 are as presented in Table 1.7.

Table 1.6: **Examples of mergers and acquisitions adapted from Bjornsen (2003)**

Consolidation by merger or acquisition

	Year						
	1996	*1997*	*1998*	*1999*	*2000*	*2001*	*2002*
Carnival		Buys Costa (in part)	Buys Cunard	Buys Seabourne	Buys Costa (100%)		Merger P&O Princess
RCCL		Merger Celebrity					
P&O Princess					Buys Aida		Merger with Carnival
Kloster	Renamed NCL			Buys Orient Line	Bought by Star Cruises		
Cunard			Bought by Carnival				
Chandris/ Celebrity		Merger RCCI					
Costa		Bought by Carnival					
Epirotiki		Forms Royal Olympic with Sun Cruises	Majority bought by Louis Cruises				

Cruise Brands

Cruise company brands, such as Costa Cruises, Orient Lines, Celebrity Cruises, Princess Cruises, Saga Cruises, and Norwegian Cruise Lines, guard their reputations with great care. As Moutinho (2000) states, branding for tourism organizations is perceived to present significant strategic advantages. Thus, a brand name can hold connotations about a corporation or a company or a cruise ship. The brand may be more

Table 1.7: **North American cruise by 10 brands (Anon, 2003b)**

Cruise Customer Statistics (market share second quarter 2003)

Cruise Line	*Customers (in 000s)*	*% of total customers*
Carnival Cruise Line	723	36.2
Royal Caribbean International	510	25.5
Norwegian Cruise Line	186	9.3
Princess Cruises	156	7.8
Celebrity Cruises	153	7.6
Holland America Line	141	7.0
Disney Cruise Line	101	5.1
Cunard Line	13	0.6
Costa Cruises	8	0.4
Crystal Cruises	7	0.3
Total	1,998	100

than a name in that a design or symbol can also be selected to represent the brand values: P&O Cruises uses a familiar nautical flag. In addition, historical events add a recognition factor to some "famous" names. Examples include the "White Star" brand that is utilized by Cunard to identify their training provision on board the "White Star Academy" (see Chapter 10 for more information).

Cruise ships lend themselves very well to the process of branding. Passengers engage with the "product," or the cruise, in a series of complex ways that enhance the opportunity to develop brand loyalty. This process extends throughout the cruise experience, from considering the glossy images on a cruise brochure and booking a cruise to the point of embarkation, when the customer first experiences the scale and impressiveness of the ship in port, through to consideration of life on board and then to departure and disembarkation. This form of vacation creates a unique relationship between passenger and ship, passenger and cruise, and passenger and brand.

Branding is important for targeting new markets, engendering repeat business, highlighting brand recognition, defining a firm's strategic approach to marketing and operations, and critically, establishing loyalty (Moutinho, 2000). Laws (1997) identifies the advantages that accrue to the major corporations, in terms of resources and marketing strength, which means they can afford to underpin brand development with impactful brand awareness campaigns, focusing in turn on the specific market segment that is deemed to be the target market for specific brand identities.

Summary and Conclusion

The cruise industry is both potent and portentous. In many ways the industry reflects strengths that have emerged as a result of the relentless growth connected with globalization. Powerful corporations exist within the industry, and they have the resources to keep pace with the constant demands that are required when investing in new ships. Countries that benefit from globalization become wealthier, and as a result large swathes of their populations are better able to purchase cruise vacations. Increasingly, sustained growth means there is greater of confidence and that in turn there is increased innovation in developing cruise products. Demand coupled with positive publicity appears to create greater demand as the potential cruising population develops more of a taste for this type of vacation. The industry also benefits from repeat business and a high degree of customer loyalty.

This chapter has examined the elements of cruising to highlight critical factors connected to its current status as a significant and growing part of the tourism and leisure field. Historic trends in cruising have been reflected upon in order to examine the way that changes have occurred over time and to understand why these changes have come about. A number of cruise brands are considered so as to contrast services and styles in order to reveal different types of cruises. This allows further discussion to be developed about the cruise brands themselves so as to encourage an understanding of the nature of this complex market.

Glossary

Cruise: A vacation involving a voyage by sea, on a lake, or on a river.
Destination: Point of arrival for a traveler or tourist.
Galley: Name for a kitchen at sea.
GRT—Gross Registered Tonnage: Relates to the size of the ship.
Keel: Lowest point of a ship's structure. This is often a beam (or plates) that extends from one end of the Ship to the other and forms the shape of the underside of the vessel.
Line voyage: Origin of the term "liner," indicating the line travelled in a journey between two specific points such as Southampton to New York.
Port agent: A professional individual or company offering local management services in ports of call for visiting vessels.

Refit: Updating of a ship's technical equipment, changes to the external or internal appearance or replacement of worn out furnishings and fittings. This is often undertaken in a dry dock facility, where the ship can be fully inspected and repairs made, if necessary.
Tug: A high-powered workhorse of a ship used to tow vessels.

Chapter Review Questions

1. Which famous cruise brand is identified as being the first to start cruise holidays?
2. Which corporation is the largest cruise operator in the world?
3. What does the acronym GRT stand for?
4. How do you calculate space ratios for passengers?
5. What parts do brand and brand image play for cruise companies?
6. What are the brand images connected to Princess Cruises and P&O Cruises?

Additional Reading

Dickinson, R., and Vladimir, A. (1997), *Selling the sea: An inside look at the cruise industry.* New York: Wiley.
Douglas, N., and Douglas, N. (2004), *The cruise experience: Global and regional issues in cruising.* Frenchs Forest, Australia: Pearson Education.
Ward, D. (2001), *Complete guide to cruising and cruise ships 2002.* London: Berlitz Publishing.

Additional Sources

CLIA—Cruise Line International Association: http://www.cruising.org/
ICCL—International Council of Cruise Liners: http://www.iccl.org/whoweare/index.cfm
IMO—The International Maritime Organization: http://www.imo.org/index.htm
PSA—Passenger Shipping Association (UK): http://www.the-psa.co.uk/

References

Anon. (2003, Oct/Nov 2003), Top 10 lines see numbers climb 9% in the first half. Lloyd's Cruise International.
Anon. (2005, 15 March), Full steam ahead, from http://www.cruiseindustrynews.com
Bjornsen, P. (2003), The growth of the market and global competition in the cruise industry. Paper presented at the Cruise and Ferry Conference, Earls Court, London.
Bond, M. (2004), Editorial, Seatrade Cruise Review Quarterly.
Carnival (2005, January 7 2005), Carnival expected to carry record 500,000 kids in 2004, from http://www.carnival.com/CMS/Articles/kidsvirtual.aspx
Cartwright, R., and Baird, C. (1999), *The development and growth of the cruise industry.* Oxford: Butterworth Heinemann.
CLIA (2005, March 16), CLIA Cruise Lines ride the wave of unprecedented growth, from http://www.cruising.org/CruiseNews/news.cfm?NID=196
Cruisemates (2005), *Queen Mary* 2. Retrieved 21 March 2005, from http://www.cruisemates.com/articles/reviews/cunard/qm2.cfm
Dawson, P. (2000), *Cruise ships: An evolution in design.* London: Conway Maritime Press.
Day, C., and McRae, K. (Eds.) (2001), *Cruise Guide to Europe and the Mediterranean.* London: Dorling Kinderlsey.
Dickinson, R., and Vladimir, A. (1997), *Selling the sea: An inside look at the cruise industry.* New York: Wiley.

Dingle, D. (2003), Cruising in the 21st Century—New developments, Paper presented at the Cruise and Ferry Conference 2003, Earls Court London.

Douglas, N., and Douglas, N. (2004), *The cruise experience: Global and regional issues in cruising.* Frenchs Forest, Australia: Pearson Education.

Ebersold, W. B. (2004), *Cruise Industry in Figures*. Washington: US Department of Transport.

Gibson, P. (2003), Learning, culture, curriculum and college: A social anthropology, Unpublished PhD, University of Exeter, Exeter.

Kontes, T. C. (2003), The cruise industry revolution. Paper presented at the Cruise and Ferry Conference 2003, Earls Court, London.

Laws, E. (1997), *Managing Packaged Tourism*. London: International Thomson Business Press.

Lutgens, F. (1992), *Essentials of Geology*. New York: MacMillan.

Michaelides, M. (2003), The latest developments in the Mediterranean Cruise Market; Challenges for the future. Paper presented at the Cruise and Ferry Conference 2003, Earls Court, London.

Moutinho, L. (Ed.) (2000), *Strategic Management in Tourism*. Wallingford: CABI Publishing.

Ocean Village (2003), webpage, from http://www.oceanvillageholidays.co.uk/

ResidenSea (2003), webpage, from www.residensea.com

Showker, K., and Sehlinger, R. (2002), *The Unofficial Guide to Cruises 2003*. New York: Wiley Publishing.

Synnove Bye, A. (2003), The future of cruise ships: the experience from the World of ResidenSea, Paper presented at the Cruise and Ferry Conference 2003, Earls Court, London.

Ward, D. (2001), *Complete guide to cruising and cruise ships 2002*. London: Berlitz Publishing.

Windstar (2003), website, from http://www.windstarcruises.com/indexcontent.asp

2

Selling Cruises and Cruise Products

Learning Objectives

By the end of the chapter the reader should be able to:

- Reflect on marketing for the cruise industry
- Consider the formulation of a range of products and services
- Analyze how cruise brands differentiate their products, develop service standards, and create brand values

The Market

A market can be described as a "system comprising two sides" (Evans et al., 2003, p. 120), with the "sides" being demand and supply. The cruise market can be further defined, according to common interpretations, in three ways: focused on the product, on satisfying a need, or on passenger identity (Evans et al., 2003). "Product focused" companies have advantages in terms of developing economies of scale, although they may fail to take account of changes that occur within their target market incrementally over time. "Need satisfaction" companies are good at understanding their customers but can have problems in making a strategic decision to identify specific focus. "Passenger identity" companies can target specific groups of passengers. Evans et al. (2003) note that most companies combine definitions in order to derive strengths from each of the three approaches.

Knowles, Diamantis, and Bey El-Mourhabi (2004a) describe world events such as the terrorist attack on New York in 2001, the subsequent Gulf War, and the prevailing economic conditions in the United States and other major countries at that time as vital in shaping the fortunes of tourism and leisure providers. Add continuing international unease in the face of potential acts of terrorism, the apparent switch to cruising undertaken by customers as a reaction to risk assessment, cruise companies' strategic decisions to facilitate easy travel to port, and the construction of safe itineraries (International Council of Cruise Lines, 2004a), and a picture emerges of an industry acting to take advantage of market opportunities.

Cruise companies target specific markets and tailor their products and services accordingly (Knowles et al., 2004a). Getting the marketing mix right is, for marketers in the cruise business, a case of building on the traditional 4 Ps of price, product, place, and promotion to include the three additional service-oriented components: people, physical evidence, and process (Aaker, 2001). Throughout this book, evidence is provided to enable the reader to deconstruct these components so the traditional or extended marketing mix can be thoroughly examined. This chapter describes the type of input that is considered by the stakeholders (cruise operators and travel agents) in the selling of cruises and then considers cruise products and the overall cruise market.

The Cruise Operators

As previously discussed, cruise operators or brands dominate the cruise market (Berger, 2004). They either own or lease cruise ships and produce the planned itinerary and cruise product so as to target specific market segments. Cruise operators can be seen as wholesalers and travel agents as retailers or brokers (Dickinson and Vladimir, 1997). However, in common with many wholesale operations, better profit margins or more attractive selling prices may be achieved if the product can be sold directly to the consumer. Therefore, the majority of cruise operators also sell their products directly to the public, acting as both cruise wholesalers and retailers. Products are developed and packaged using market research, negotiation, and sales and marketing.

All cruise companies exert considerable effort to establish brand values and in constructing cruise products that are designed to meet and, ideally, to exceed passenger expectations. Market research collects data from existing and potential passengers using a variety of research techniques. This data can be used to interpret customer behavior and predict buyer responses to new products. Increasingly, anthropologists are employed to study target groups or individuals so that companies better understand why people act as they do.

The products developed are an amalgam of services and facilities, some of which generate revenue while others are included at no additional cost. This means that most cruises have fixed costs relating to elements such as transportation (fuel), food, labor, port administration, and customs and have variable costs relating to other elements such as beverages or shore excursions. The cruise operator aims to reduce costs as much as possible without negatively affecting quality. Negotiation is done to achieve the best ratio of price to quality and to take advantage of economies of scale and negotiating power. Negotiation is therefore undertaken for a range of consumables, from engine or deck department stores through to hotel department stores. In terms of buying power, considerable advantage is accrued by the largest corporations through such negotiations.

Traditionally, cruise companies have used travel agents as a primary distribution channel while concomitantly selling directly. Irrespective of the mode of distribution and despite the growing importance of the Internet as both a distribution and marketing tool, cruise companies rely on the cruise brochure to sell cruises. Vellas and Becherel (1995) describe how tourism operators design brochures with color images, a carefully planned layout and promotions designed to attract early booking. They note that the ratio of brochures to sales can be between 10 and 30 brochures for one sale. Pricing strategies are carefully considered to encourage early action by promising discounts for early booking. Low season pricing is adjusted to appear less costly than high season pricing. Lead-in prices relate to basic cabin accommodation, and supplements are payable for attractive alternatives such as outside cabins or sea view cabins with balconies. Premium products, such as suites with butler service, come with premium pricing.

Brochures are produced well in advance of the cruise date and planning has to take into account fluctuating prices, rates of exchange for items purchased outside the country of the operator, and changing market conditions (Dickinson and Vladimir, 1997). It is reasonably common for cruise operators to update brochures so as to react to changing conditions. Changes can include offering different prices in later editions and, in some cases, making amendments to the product if the change has come about for reasons outside the cruise operator's control.

The Internet is used predominantly as a complementary marketing tool. A website can be a point where information is presented to potential and actual customers to help them to find out more about the cruise package in a way that brochures never can (Berger, 2004). For example, customers can visit the passenger feedback pages to see what passengers are saying about their vacations and customers can follow links to other important information—for example, finding out about immigration or health matters overseas. The Internet can also be used to enable clients to book online. In this mode, the customer assists the cruise operator by providing data in a format that can be easily manipulated and also helps cut out the costs associated with booking through a travel agent or sales assistant. The Internet can also be used to capture data for immigration purposes and for financial control, thus potentially saving on administration costs.

The Travel Agent

The travel agent's core purpose is selling tourism products for commission. Most travel agents belong to professional associations that guarantee clients protection if the travel agent has serious financial problems. The American Society of Travel Agents (ASTA) and the Association of British Travel Agents (ABTA) are typical of such associations. Travel agents sell travel products such as airline tickets and tourist packages. They can also arrange insurance, car hire, and hotel accommodation.

However, the traditional travel agent is changing (Hatton, 2004). Faced with ever increasing competition from Internet intermediaries or online agencies, travel agents are finding themselves operating in a volatile marketplace. Airlines have cut commission rates. Travel firms have been aggressive in selling directly to clients, thereby cutting the travel agent out of the distribution system. Hatton notes that the travel agent's strength, providing a highly personal and personalized service, also undermined their status and led travel companies to try to nurture brand loyalty by developing relationships with the client directly. In response, Hatton (2004) highlights the need for agents to accept the changing realities and to work closely with travel companies to develop in-depth product knowledge and retain customer loyalty by being efficient in what they do.

Examples can be seen in the context of cruise vacations. Some travel agents specialize in the cruise industry, forming alliances with cruise brands to focus on selling their product. In these circumstances, travel agents receive high levels of support from the cruise operators, who provide specialist sales events, training for the sales agents (including orientation cruises), and customized marketing materials. The Passenger Shipping Association (PSA) has established the PSA Retail Agents (PSARA) scheme in the UK with the primary objective of increasing sales through customized product training and dissemination of information to accredited retail travel agents (PSARA, 2005).

Marketing Actions and Alliances

A market is the place where sellers and buyers meet to do business (Evans et al., 2003). Marketing as a sophisticated process and discipline has emerged from this basic premise. Accordingly, cruise operators develop their product to meet customer expectations and then design a plan for marketing the product. The plan can include advertising, promotion, merchandising, and public relations (PR). Advertising uses a number of communications media including: commercials on radio, television, the cinema, the Internet, newspaper and magazines, posters, and billboards. PR can be channelled through editorial or features in travel publications, newspapers, and magazines. Merchandising reinforces the brand by using items such as pens, desk pads, or other mementos to remind users about their vacation. Promotion can be associated with advertising or it can be incorporated into events for visiting groups on board cruise ships.

Cruise operators may form strategic marketing alliances with other service providers to create synergies or provide customers with incentives for remaining loyal. Crystal Cruises is a member of the Luxury Alliance, which includes Silverseas Cruises, Orient-Express Trains and Cruises, Leading Hotels of the World, and Relais and Châteaux (Luxury Alliance, 2005). The "World's Leading Cruise Lines" alliance includes Carnival Cruises, Holland America, Cunard Line, Seabourne, Costa Cruises, Princess Cruises, and Windstar and provides incentives for loyalty within these brands (World's Leading Cruise Lines, 2005).

Loyalty

Loyalty is important both for the client who enjoys acting as an ambassador for the brand and for the company that values retaining such a customer. In effect, this type of client works for the brand by spreading positive comments about the cruise to friends and acquaintances and, as a result, is an important part of the marketing equation. Companies such as Princess Cruises operate an incentive group called the "Captain's Circle." This club has three levels: Gold (2 to 5 cruises), Platinum (6 to 15 cruises) and Elite (16 cruises and more). The benefits include priority discounts, special events on board, preferential services, and other benefits depending on the level of membership (Princess Cruises, 2003).

The Cruise Product

Like some other tourism products (Vellas and Becherel, 1995), the cruise has three economic features: heterogeneity (the product possesses a broad mix of variable components that render the experience unique for the individual tourist), inelasticity (a cruise product is "perishable" because it cannot be stored if it is not sold) and complementarity (the cruise product is not one single service but a series of complementary services that when taken together form the cruise experience).

The cruise is a defined package that may include travel to the port of embarkation, an itinerary spanning a defined period of time, an element of inclusive services and facilities (such as meals, entertainment, and leisure areas), accommodation to a specified standard, and various other services that are available at an extra charge. The inclusive nature of the package will depend on the pricing strategy of the cruise operator. Some operators offer cruise-and-stay or cruise-and-tour packages that include an additional element at the beginning or end of the cruise in the form of a tour of an area or a defined number of nights in a resort hotel. The following elements portray the products of cruising.

Accommodation

For many passengers, the choice of accommodation seems to be simply a matter of identifying the price that is acceptable relative to the standard of accommodation available. However, a glance at the pricing structures operated by cruise companies quickly reveals that selecting accommodation is more complex than would first appear to be the case. Some cruise companies refer to the accommodation as cabins, but terms such as staterooms, minisuites, and suites are frequently used to replace or complement this nautical term. Some cruise companies sell penthouse suites on board their vessels, and these tend to be the largest, most luxurious, and most expensive options (Mancini, 2000).

Although cabin sizes can vary from just under 11 square meters (120 square feet) to over 85 square metres (900 square feet), the norm tends to be approximately 18 to 23 square meters (200 to 250 square feet). Even a cabin of approximately 14 square meters (150 square feet) is likely to have four beds configured as lower and upper berths. The upper berths can be folded back to create more space or to cater to two passengers rather than the maximum of four. Cabins may also permit the lower berths to be moved together to form a large queen- or king-size bed. The largest cabins can be configured as suites with a lounge area. All cabins on modern cruise ships tend to be *ensuite*; that is, they have a shower, room and toilet or a bath, shower and toilet (Dervaes, 2003).

Generally, cabins are compact versions of the equivalent hotel bedroom accommodation. The storage areas are carefully designed to maximize the use of space so as to create an impression of a very efficient, customized facility. Space is typically at a premium on cruise ships, and vessels are constructed to maximize the area that can generate revenue. Because of this design constraint, some cabins will inevitably be preferable to others. They may have a good view or a restricted view, a location that is perceived to be more or less appealing because of proximity to certain facilities. For example, if a cabin to located close to elevators or other sound-generating elements, some passengers may be unhappy about the resulting background noise. Ships are carefully designed to minimize noise, yet, on most vessels (as in hotels ashore), some cabins are recognized as being potentially problematic.

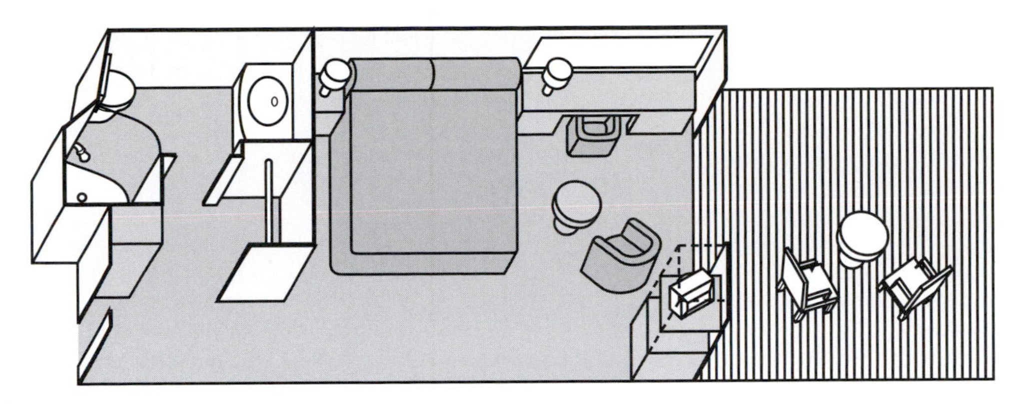

Figure 2.1: Diagram of a stateroom

Customers learn how to make decisions from the brochure or cruise brand website. These information sources provide a variety of data that is intended to help the customer make a selection. Floor plans provide a miniaturized cut-away view of the cabin (usually produced to reflect three dimensions; see Figure 2.1) that shows the placement of the furnishings, main features, typical layout, and *ensuite* facilities. Generally, photographs are used in conjunction with the floor plans or, in the case of the website, a 360° scanning view can be shown. A description of the cabin contents usually accompanies these images.

The most common way of identifying cabin locations is by using deck plans. These are representational ship's plans that, when viewed in conjunction with a cross-section diagram of the ship, help customers to identify the precise location of a cabin or facility on board. These plans are unique to each vessel, although similar vessels may well have many common features. The deck plans are produced in color so that a code can be used to identify cabins by type and, therefore, by cost (see Figure 2.2). On some vessels, cabins that are on lower decks are less expensive while cabins that are on higher decks are more expensive. However, this pattern is not reliable for all ships. It is possible on a deck plan to identify cabins as follows:

- Inside cabins or staterooms. These cabins lack natural light, although use of ventilation, air conditioning, mirrors, and artificial light frequently disguises this fact. Inside cabins tend to be the least expensive accommodation on offer.
- Outside cabins or staterooms. These cabins will have a porthole or a window. Most modern cabins tend to have larger picture windows.
- Outside cabins or staterooms with a veranda or balcony. As cruising develops, more accommodation is being produced to include private verandas or balconies with more private space.
- Penthouse suites or suites that may be with or without a veranda or balcony. These cabins tend to be the most expensive accommodation on offer.
- Cabins or staterooms with additional beds (berths).
- Cabins or staterooms with interconnecting doors.
- Cabins or staterooms or suites with facilities that are appropriate for people with disabilities.
- Cabins or staterooms with either shower or bath.
- The proximity to facilities, lifts and location compared to other decks and cabins.
- The proximity to safety equipment such as lifeboats, which may obscure the view from a picture window.

It is suggested that passengers expect more from cabin accommodation or staterooms than was previously the case. In many respects a cruise brand, carrying thousands of passengers and crew on large ships that offer a broad spectrum of leisure activities, can provide a balance for those seeking enhanced levels of privacy by offering more spacious cabins that have attractive features such as balconies.

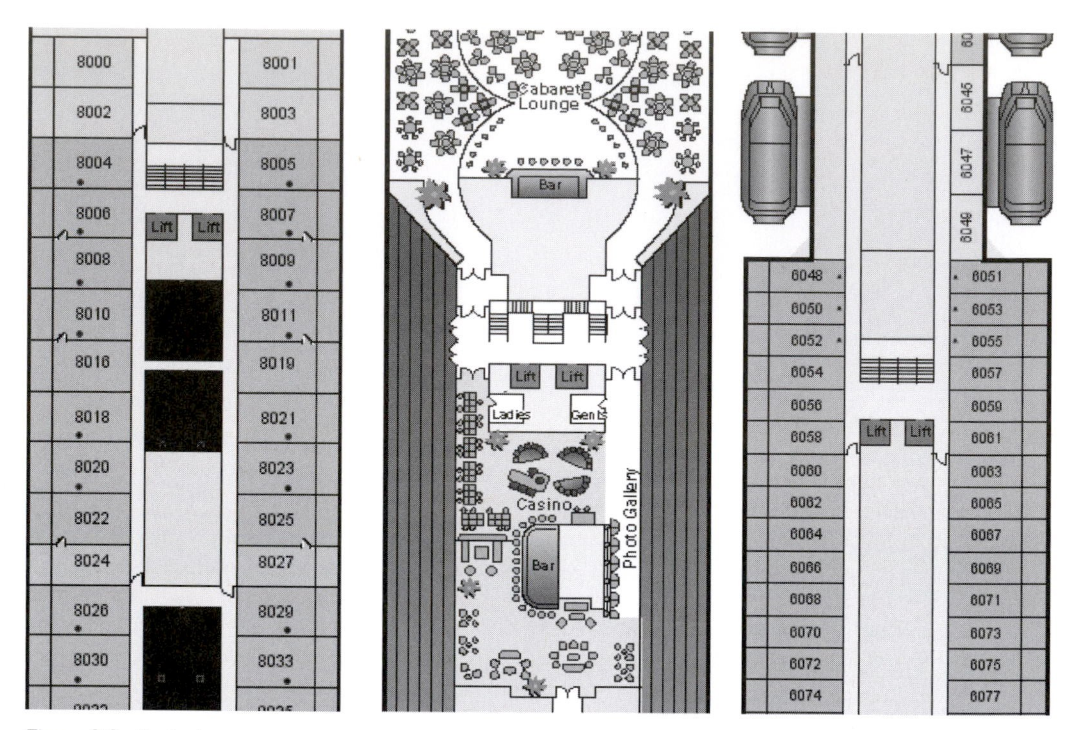

Figure 2.2: Deck plans

Using a deck plan, customers can select the accommodation to a precise degree. Doing so can benefit repeat passengers who have distinct preferences or satisfy passengers who have exacting requirements. However, a problem can occur because of what appears to be a commonly held misconception about upgrading. Some travel agents are reported to advise passengers that when they get on board the expected procedure is to complain in order to get a higher category of accommodation without paying more. Sales and reservation teams aim for 100% capacity, and the flexibility on board is severely constrained by the occupancies achieved through this approach. Spare cabins are scarce resources that are vital for dealing with problems that may arise, such as plumbing faults or electrical failure, and as a result upgrades are virtually impossible.

By producing deck plans, the cruise company can help prospective clients to identify the cabin location of their choice. Also, in a more practical sense, they also assist the passenger when he or she boards the vessel to be oriented more easily into shipboard life. Deck plans are reproduced in a fold-down version to act as an easy reference map for passengers on board.

Cabin facilities vary depending on the cruise brand, but a basic cabin or stateroom is likely to include the following:

- Two single beds that can convert to a queen- or king-size bed.
- Optionally, an extra two upper berths that can recess into the internal wall and are reached by store-away ladder.
- Bedside tables.
- A vanity unit and writing desk with built-in drawers, storage areas, mirrors, and chair.
- An additional small table and chair.
- Television and radio with remote control. The range of programming can include films and an onboard TV channel.

Tea- and coffee-making facilities.
Safe, hairdryer, refrigerator/minibar.

- Direct-dial telephone.
- Bathroom with shower and toilet.
- Air conditioning.

Some brands may also upgrade the facilities to provide:

- Internet terminals.
- CD/DVD/VCR and stereo.
- Balcony with furniture.
- Separate living area or lounge with suitable furnishings.
- Separate dressing area.
- Hot tub.
- Separate toilet.

The décor in passenger cabins reflects the standards associated with the brand. Soft furnishings are coordinated with carpeting and artwork to create the desired ambiance. Color schemes are selected to fit the mood and to create an overall harmony with décor and furnishings. Lighting is strategically located to provide the appropriate level of light for the purpose, whether reading, writing, or personal care. Mirrors are used strategically to accentuate space and light.

Passenger cabins and staterooms are attended to by cabin stewards who monitor the general condition of the accommodation and perform routine cleaning and service. The steward's daily tasks include making up beds, changing linen and towels as required, cleaning and vacuuming and ensuring that the cabin is prepared to a prescribed specification (for example, the bathroom is laid out and the complementary items are displayed). The steward also checks the room minibar and works with supervisors to deal with technical problems. Housekeeping supervisors and managers inspect cabins to ensure that standards are maintained. Room service may be provided by the room steward or separate personnel. Many large cruise ships have what is called a bell box, where a small team of chefs and room service stewards address requests for food and drink to passenger cabins as required. Some suites and penthouse suites are allocated a butler to provide more personal service. The butler can facilitate the catering and service of parties and can arrange for other services and products that may be requested by the passenger.

Dining on Board

The archetypal view of a cruise ship as a place to indulge in good food, good wine, and good company is as true today as it was in the heyday of the traditional liners. Food is perceived to be a significant element of the cruise product. For most passengers, the cost of eating on board is included in the price of the voyage. There are exceptions such as restaurants, which carry a supplementary charge, but in the main the inclusive nature and the high customer expectation of the dining experience are a fundamental issues. Most cruise brands aim to differentiate what they do through the provision of food and dining options. They can create opportunities to define the product and to differentiate the brand by constructing menus with a focus on a particular style of cuisine and by designing restaurant and dining areas with a particular décor and atmosphere in mind. Therefore, on certain ships the restaurant may have an Asian theme with Japanese cuisine and on others there may be an Italian theme. Cruise dining can be a densely calorific affair, yet corporate chefs take great care to meet specific dietary needs when they design menus.

Some brands have introduced greater options for personal choice in dining. By moving away from traditional dining arrangements that offered two sittings at dinner supplemented by open sitting at breakfast and lunch in large dining rooms, these companies were able to change the formula and attract clients who wanted more flexibility. Most large cruise ships operate at least two large 500-plus-seat restaurants that sit to either side of a galley with a double-ended servery or hotplate. (The servery or hotplate is the area in the galley where waiters collect food for service.) This arrangement facilitates the service of large numbers of people at dinner without the creation of lines at the door. Linking dinner service sittings with entertainment schedules ensures that passengers are not left disgruntled at having to go to one sitting or another. Coordinating breakfast and lunch is usually less of an issue because passengers have alternatives such as buffet or room service breakfast and buffet lunch. Because the ship tends to visit ports during the day that also has a secondary effect for producing and serving breakfast and lunch.

The Buffet

The buffet is a flexible option on a cruise ship. Often it is located on an upper deck and frequently it is designed to extend from one side of the ship to the other with each side being a mirror image of the other. This setup enables large numbers of passengers to move through without creating bottlenecks. At quiet times and when service changes from one meal to another, one side of the operation can be closed for cleaning and changing or replenishing the food items. This practice creates a 24-hour facility that is both flexible and economical.

A small team of chefs under the supervision of a sous chef services the buffet. The galley team is supported by buffet assistants and supervisors who help customers, clear tables, and serve drinks as required. The buffet employs equipment that is designed to present food items at the correct temperature and in the best way to make that food attractive to passengers. Main food items such as soups, meat and fish entrees, cold dishes, and desserts are changed daily according to the duration of the cruise itinerary, although some standard items such as breads, salad items, dressings, and condiments are available daily. Food items are designed with a culinary nod to each approaching port on the itinerary.

The buffet (see Figure 2.3) requires fewer staff than the traditional restaurant, and because of the way it can be operated with simple table layouts, standardized areas, for beverages, fewer carpeted

Figure 2.3: Buffet servery

areas, and large picture windows overlooking either the sea or the port, the servicing routines are more easily carried out. The buffet takes the strain off the restaurant at breakfast and lunch, thus allowing staff to be deployed more effectively and for the galley to plan production more accurately. Very little waste is created because food can be carefully produced in reaction to prior patterns of demand and prevailing consumption. Buffets are frequently organized with the galley and servery in the middle of the room and tables and chairs around the outside beside the windows. Buffets tend to be designed with a wash-up area beside the galley. Tables are cleared to a collection point (sometimes referred to as a DJ's box). Dirty plates and food residue is then taken by cart to the dish washer. Food and supplies are transported in the specially allocated lifts from the main galley to the buffet galley.

The Main Restaurant

Passengers can eat as much or as little as they wish, and nowhere is that more evident than in the main restaurants. While passengers may stack their plates full in the buffet, in the main restaurant, food is conveyed to the diner as frequently as the diner requests. Diners who may be concerned about what others might think as they are seen carrying a pile of food from the buffet seem to believe that becomes obscured if quantity is disguised within the routine of ordering courses from a menu.

In the main restaurant or restaurants, a menu is produced to reflect and to differentiate the brand. The way the menu is configured might lean toward acknowledging passengers used to eating out in the United States or in Italy or in the United Kingdom. This can involve the provision of distinct courses, the names of the types of courses, the food items included within the courses, and the language adopted in describing the food items and courses.

The main restaurants tend to reflect a style and standard that is redolent of a more formal dining experience, with the use of uniforms to identify the maître d' (the abbreviated version of maître d'hôtel, meaning the overall restaurant manager), head waiters, waiters, and assistant waiters or commis waiters, and the presence of professionals such as sommeliers or wine waiters. The combination of white tablecloths, sparkling cutlery and glassware, careful selection of colors and hues, materials and furniture, and subdued lighting add to the effect, as do the provision of music (sometimes live) and the theatricality of the environment. The setting is important in getting passengers to interact with each other and with staff. The food and wine are the reason for being in the restaurant, but the experience is enhanced by the social factors.

Service styles vary depending on the brand and passenger expectation. Full silver service may be adopted by the luxury brands, semi-silver or plated service by the contemporary and premium brands, and budget brands may have a combination of buffet and plated service. Each style of service is correlated to the skills of the server and the ratio of staff to passengers. Full silver service requires the greatest degree of skill in serving and presenting food, and as a result there is a need for a lower ratio of staff to customers. Whichever service style is used, the common denominator for service staff is the need to develop the appropriate level of interpersonal skills. Table sizes vary, with tables available for between two and eight people. Larger tables are more common on two-sitting dining plans. Many cruise brands are introducing free dining situations where customers can prebook tables for certain times and ask to dine privately or join a group.

The formality of the dining area is not without reason. Formality and dress codes are features on many cruises. While new brands place an emphasis on informality, the norm for most cruise brands is still toward creating opportunities for passengers to dress to impress. On these cruises, passengers have the opportunity to dress formally once every 4 or 5 days.

Other Dining Options

While most meals are regarded as a composite part of the cruise experience and inclusive with the cost of the vacation, increasingly, cruise brands opt to generate revenue by providing more choices that add value to the overall dining experience. For example, on the '*Star Princess*', passengers can reserve a table in a variety of alternative restaurants, such as Sabatini's, an upscale Italian restaurant, or the Tex-Mex grill, both of which cost extra. On '*Ocean Village*' the Bistro is a restaurant that has food produced to the specification of a leading UK-based chef, and this option is also charged at a supplement. Al fresco dining is offered by some cruise brands, creating an opportunity for passengers to eat under the stars (Princess Cruises, 2003).

Other restaurants such as pizzerias and burger and hot dog grills create alternative options that might be seen by certain of passengers, such as children, as more appealing than the formal setting of the restaurants. Ice cream stands may also be located on decks close to swimming pools and sunbathing or leisure areas. Afternoon tea and, in some cases, high tea (for young families) are served from restaurants and buffets. Finally, passengers make use of room service if the choice of dining options really does not meet their needs.

Bars

Generally, most bars begin to get busy after dinner. The routines of sailing are established relatively quickly as passengers find their way around and work out what they want to do and where they want to go. The busy times are from around 22.00 (10:00 pm) onward, but there are numerous opportunities for passengers to purchase drinks and for bars to generate revenue:

Sailing day: Drinks are available on upper decks as the ship departs from port. Working from bars at key points, bar waiters mingle among the passengers selling cocktails and drinks to celebrate the departure. Live music is played to add to the atmosphere.

Pre-ordering dinner: A wine preordering point is usually made available so that customers can ensure they order the wine that they want for dinner. During dinner, wine and beverages are available from a dispense bar usually located within the galley. Lists for wine, liqueur, Cognac, and fine whiskies, and liqueur trolleys and merchandising displays all support the sales initiative. Sommeliers and wine waiters are on hand to help passengers.

Theater: Table service is available in all the entertainment venues. Cocktails of the day and special promotions are offered to highlight the range of options available.

Bars: Various bars are targeted at particular groups of passengers, such as sports bars with sporting memorabilia and recorded or live sports displayed on television screens or traditional lounge bars that use dark wood and comfortable settees and chairs to give a "club" feel.

Champagne and caviar bars appeal to a certain clientele and exude quality and exclusivity. Piano bars combine relaxed intimacy and friendly ambience.

Nightclub: These venues generally have table service and a cocktail menu. Depending on the clientele on board, different products are likely to be available. Cocktails are popular on cruise ships, and many bars use premixed blends that must to be combined with a spirit and ice before being shaken or blended and garnished.

During the day: Drinks can be purchased from at least one bar inside the vessel and from pool bars on the sun decks throughout the day. Drinks are also available from mobile points in the buffet and in the restaurant when meals are being served.

Lounges: Passengers congregate in a variety of places for quiet moments or to play cards or read a book. While bar service is an option, tea and coffee are more likely to be consumed. Ships often develop a range of lounges according to the needs of passengers. These areas might include a library, a card or bridge room, a writing room, an observation lounge, or general lounges. These areas can be used for quizzes and competitions, wine tastings, and small group meetings.

The various bars (see Figure 2.4) can also be used for a range of activities including art auctions, competitions, karaoke, dance classes, fashion shows, and other entertainment. The bar staff work through a rotation that covers the various areas within the ship and creates a fair and equitable pattern of work for everyone.

Figure 2.4: Bar on '*Aurora*'

Entertainment

The entertainment staff works for the cruise director, who in turn reports to the hotel services or passenger services director. This element of the cruise product does not generally create additional revenue, although sales arising from entertainment activities can be made indirectly.

Theaters are the venue for the headline activities such as musical extravaganzas, comedy clubs, cabaret, or magic shows. The theaters are also the largest gathering areas for passengers, so they can also be used for emergency drills and as a meeting point for shore excursions. There are usually two or three shows each evening. Shows and performances operate on a rotating schedule, which is designed to ensure that the program appears fresh, interesting, and new.

The daytime activity program is produced by the entertainment staff. Events during the day are published in the ship's newspaper. These events can be very diverse to suit the types of passengers on board. The team also includes port lecturers, dance instructors, and lecturers for cyber cafes or IT suites. The entertainment staff can manage fashion shows, arts and crafts demonstrations, culinary demonstrations, and wine tastings, often working with staff from other departments on board.

Musicians are employed to provide support for theatrical productions, show bars and bar areas, sailing days, deck parties, and piano bars. A technical team provides cinema support, IT support for computers on board, stage support for lighting, sound, and special effects. They also are available to help the musicians if they require technical support.

The leisure staff provides support for sporting activities such as onboard golf, and various water sports such as jet skis, water skiing, scuba, and windsurfing that may be available from the aft section of some cruise ships. The vessel may rent bicycles to passengers to take ashore. Fitness classes such as aerobics, Pilates, and yoga are operated within fitness suites. A separate team can design activities for children, noting the specific needs that relate to different ages. Princess Cruises operates the "Pelican Club" for 3- to 7-year-olds, the "Pirateers" for children aged 8 to 12, and "Off Limits" for teenagers 13 to 17 (Princess Cruises, 2003).

Shore Excursions

Shore excursions are sold before and during the cruise. They generate revenue but are also designed to add value to the cruise experience. Because of the constraints on time, shore excursions or tours ashore are configured to maximize the experience for passengers. The range of transport options can be vast, depending on the port of call, and can include traveling on launches, by coach, by bicycle, by horse-drawn buggy, or taking a helicopter trip. Booking through the cruise company provides certain advantages. For example, if something were to go wrong such as a vehicle breakdown, the cruise company would take full responsibility to sort the problem out and ensure the passenger was not overly inconvenienced.

Shore excursions often use third-party tour operators to provide tours as well as a network of contacts to develop their shore excursion program (some cruise companies also own tour operations and can take advantage of this fact). The organization of tours for passengers is like a military operation involving planning, crowd control, careful timing, and efficient communication.

Beauty, Therapy, and Hair Care

This area is also generates revenue. Some cruise brands contract the service as a concession (an arrangement where the operator comes to a financial agreement with the cruise company to operate on board) and others employ their own staff directly. A number of well know beauty or hair stylist brands such as Steiners operate on cruise ships and there are some brands, such as Lotus Spa, that are created to uniquely identify the style of operation that is run on board specific cruise ships. Increasingly, cruise ships recognize the growth in "well-being" or "spas" as contemporary lifestyle choices.

Treatments available include chakra stone therapy, thalassotherapy, foot massage, manicures and pedicures, hair styling, oxygenating facials, body wraps, and health and nutrition lectures.

Shops

Shops on board provide a welcome indulgence for passengers seeking to enhance their usual routine of retail therapy. Just because they are at sea does not mean they cannot browse and pick up items of interest, or in some cases, necessity. Indeed, an added benefit that attracts shoppers is that the goods are sold as duty free. Ships traveling in international waters do not generally pay duty. As with beauty therapy, shops on board can be either concessionary or operated directly. If they are operated directly they tend to be line managed by the staff purser administration or someone of similar rank. The range of shops can include a jeweler, fashion stores for women and men, a gift shop, and a more general store that may also sell alcohol and cigarettes.

Shops on board usually occupy a central area within the ship that mimics the shopping mall of a large city. The trend to construct a large impressive atrium on megacruisers suits this tendency and creates an additional advantage in allowing shops to develop temporary market stall areas by moving into the spaces opposite and adjacent to the shops' main locations. This practice increases the overall trading area and helps to create a bustling market feel. Shops operating under concession are managed by companies such as Miami Cruiseline Holdings, Harding Brothers Duty Free, Nuance Global Ships, and Flagship Retail Services Incorporated.

Photography

The ship's photographers are kept busy in the endless cycle of capturing magic moments. The opportunities to record important events occur from the point of embarkation right through to departure from the last port of call. This ensures that passengers can purchase posed, professional pictures in special presentation packs and have something special to remember. Contemporary cruise companies have invested heavily in digital technology so as to customize photographs with digitally composed and mastered backdrops that are relevant to the port of call or event. In this way, passengers photographed disembarking in Venice will have their photograph framed within a montage of Venetian images as appropriate. Photographers appear at the gangway when passengers arrive onboard and are present during cocktail parties, gala dinners, and formal events. They accompany tours and attend passenger meetings. Their job is to get the picture and then to sell the picture to the passenger.

Pictures are presented in corridor display areas so as to be easily viewed by passengers who may be on their way from a restaurant to a show bar. It is difficult for passengers not to stop and look, and the sale can be confirmed with the application of carefully considered sales techniques. Some photographers are employed directly by cruise brands, and others are contracted by concessionary operators such as the Cruise Ship Picture Company, Image Photo Services, Inc., Ocean Images, Ltd., and Digital Seas Internet Cafes.

Casino

Casinos on board seem to meet the expectations of some passengers for that James Bond moment. Casinos are described as a venue for "action and excitement" (NCL); "you'll have the time of your life" in a Carnival Cruises casino and Royal Caribbean says, "There's nothing like the excitement of a winning hand at poker or a slot machine paying off." The glitz and glamor that are portrayed in Las Vegas–style casinos are emulated on board cruise ships. Gambling is perceived as a pastime for winners, and correspondingly the cruise, as a type of vacation, is synonymous with success.

Cashless ships are becoming commonplace in the cruise industry. Passengers receive a card that allows them to purchase goods on-board and credit that to their account. Casinos also use this mode of purchase and sell tokens for slot machines or chips for gambling. Casinos are allowed to open upon sailing, although some ports permit the casino to trade even when the ship is in port. In certain jurisdictions cruise ships cannot open the casino until they are 3 miles from shore. Casinos are generally operated to strict codes. For example, the International Council of Cruise Lines (ICCL), a nonprofit trade association consisting of the 17 largest passenger cruise lines that call on major ports in the United States and abroad, publishes guidelines as seen in Table 2.1.

Casinos are open to players over the age of 18 (21 in Alaska and some other ports). Most casinos have a dress code and are operated with minimum and maximum bets posted clearly at tables. Typical games on offer can include blackjack, craps, roulette, Caribbean stud poker, three-card poker, baccarat, and video poker.

Table 2.1: **Gambling guidelines (International Council of Cruise Lines, 2005)**

The ICCL is dedicated to helping the cruise industry provide a safe, secure, fun, and entertaining ship environment for its passengers. Among the services that illustrate this commitment to fun entertainment are the gambling casinos found on most of the ICCL member vessels. Industry guidelines address the equipment, conduct of games, internal controls, and customer service for casinos on cruise ships.

The guidelines generally foster the following goals:

- To provide reasonable rules by which all gambling onboard member vessels is conducted to ensure fair and professional gaming at the highest level of integrity to the player.
- To provide internal controls upon which passengers can rely to assure them that the gambling is operated with the utmost integrity.
- To provide a form of entertainment for passengers that is responsive to the customers' requests.

Weddings, Renewal of Vows, and Other Celebrations

While on board, passengers can elect to celebrate special occasions, and on some vessels couples can get married. The ability to perform weddings does not exist on all ships because of national laws that apply to the various ships and their flags or registration. However, where the law allows, the ship's captain can perform a marriage ceremony. This creates a unique opportunity for passengers, and in response cruise companies have developed a selection of inclusive packages to cater to these market, including coordination of the entire event. The package can include champagne, photographs, the wedding reception, flowers, the ceremony, the wedding cake, and souvenir items.

Other passengers can purchase a package to renew their vows. Again, the captain presides over the event and the package can be customized to include spa treatments, champagne, and a formal ceremony. Honeymoons, anniversaries, birthdays, and other special celebrations can all be catered to as part of a package.

Brand Values and Vessel Classification

The size of the ship will have a major impact on the kind of cruise experience passengers enjoy. Large megaliners typically feature multiple swimming pools, casinos, spas, many dining options, and lots of activities. Small ships forgo some amenities in favor of focusing on a destination and a different kind of cruise experience. Cruise observers classify ships in a variety of ways, such as the number of passengers the ship holds; the quality of food, drink, and accommodation; and an overall measurement of the cruise experience. While no single standard exists, there is value in analyzing what is done to establish classifications and standards of quality. Many cruise lines operate ships in different classes so as to attract a targeted clientele. In this way a company can design specific itineraries that are commensurate with the size, the product range, the target market, and the selling price.

Classification of Scale

Table 2.2 classifies ships either by carrying capacity or distinguishing characteristic (Spartan Travel, 2005).

Classifications by Status and Value

While Spartan Travel (2005) also identifies three expense category ratings—budget, midrange, and luxury—other industry observers take a broader view in their analysis. For example, CLIA lists five categories: Luxury, Premium, Resort or Contemporary, Niche or Speciality, and Value or Traditional (CLIA, 2005c). This classification is also adopted by Bjornsen (2003).

Table 2.2: **Defining vessel types**

Definition	*Description*
Megaliner	over 2,000 passengers
Superliner	between 1,000 and 2,000 passengers
Midsize	between 400 and 1,000 passengers
Small	less than 400 passengers
Boutique	special purpose, usually less than 300 passengers
Sailing Vessel	a ship primarily powered by wind
River Barge	a ship that primarily cruises on inland rivers

An analysis of this classification suggests that ships that offer the ultimate in comfort, cuisine, and attentive service are called luxury brands. This product tends to be the most expensive, and while ships in this category are usually small, there are exceptions. The accommodation and public areas are always finely appointed and carry relatively few passengers in spacious staterooms, suites, or duplexes, which tend to have balconies. Service options may include butler service. These ships are usually the equivalent of what used to be called "five star" quality. Some brands, such as Crystal Cruises, have adopted the rubric of "six stars" to identify their unique level of quality.

Next in rank are premium brands, which offer above-average food, service, and amenities, including a high number of outside cabins with balconies. These lines aim to appeal to broad age groups by providing a diversity of attractions for children, young adults, and older adults together with wide range of entertainment. Premium brands, like luxury lines, have a high ratio of space to each passenger.

Contemporary brands are the equivalent of floating resorts with capacity spanning from the mid-size vessel to the most recent megaliner or megacruise ship. These vessels provide choice and value with a contemporary twist. On board amenities, such as an ice rink, golf range, or climbing wall, are often impressive. Style may well be casual, although opportunities still exist for passengers to dress up on optional formal evenings.

Niche or speciality cruises focus on a specific aspect of the cruise, such as the destinations, in order to develop a unique product. These types of cruise companies are specialists in their fields. They pride themselves on having expertise in aspects such as cultural interpretation, soft adventure, or enrichment activities. These cruise companies target the more experienced traveler.

Budget or value brands usually use medium-size, refurbished, older ships with fewer facilities than the new megaships. This product will often take advantage of lower staffing ratios by using, for example, self-service options for main dining events. The ships are generally classically designed, and while the products are economically priced, the options of choice and travel make this form of vacation attractive to those who are relatively new to cruising.

Summary and Conclusion

This chapter identifies a number of interrelated issues connected to sales, marketing, and the cruise industry. In examining marketing and reflecting on the current infrastructure for selling cruises, a picture emerges that presents a view of the dynamics in context. The cruise market is evolving. It is becoming multifaceted, with an emphasis on targeting and market segmentation and continually identifying opportunities for growth and new developments. Correspondingly, the cruise product is also becoming more diverse as operators continue to seek new ways of meeting passenger needs and satisfying expectations. The cruise ship in the twenty-first century still relies on people for the critical part of the service product (Evans et al., 2003) and in many ways the human element will continue to make the difference between achieving a high-quality outcome and being ordinary.

This chapter has examined marketing in general terms, selling from a wholesale and retail perspective, distribution options, and the part the Internet and loyalty programs play for cruise business. Cruise products are described to establish an overview of what is available and why it is provided. The list of products is not inclusive but an indication of the type of services that are generally provided.

Glossary

Brand: A trade name given to a product or service.
Hotplate: The place within the galley where the food server collects main food items.
Semi silver: A service style in which some items are served and others are plated.
Silver Service: A service style in which the waiter uses spoon and fork or other specialist service techniques to serve all food from a service dish to the plate.
Sommelier: The wine waiter.
Sous chef: The term *Sous* means under in French. This is generally the Executive Chef's most senior assistant.

Chapter Review Questions

Consider the cruise ship as a product and reflect on the component parts that together form the cruise experience.

1. What is the marketing mix?
2. What are the various ways that cruises are sold?
3. Why are travel agents more commonly used for selling cruises?
4. What are the options for classifying cruise ships and why can this be useful?

Additional Reading

Berlitz Guide to Cruising and Cruise Ships, by Douglas Ward
The Unofficial Guide to Cruises, by Kay Showker
The Complete Cruise Handbook, by Anne Vipond
http://www.oceancruiseguides.com/cruiselines/clclass.html
http://www.onboard.com/products_promotions.asp
http://www.geocities.com/TheTropics/Shores/5933/addresses.html#GIFT

References

Aaker, D. (2001), *Strategic market management* (6th ed.). New York: Wiley.
Berger, A. A. (2004), *Ocean travel and cruising: A cultural analysis*. New York: Haworth Hospitality Press.
CLIA (2005), Plan your cruise. Retrieved 22 March 2005, from http://www.cruising.org/planyourcruise/crsfinder.cfm
Dervaes, C. (2003), *Selling the sea* (2nd ed.). New York: Thomson.
Dickinson, R., and Vladimir, A. (1997), *Selling the sea*. New York: Wiley.
Evans, N., Campbell, D., and Stonehouse, G. (2003), *Strategic management for travel and tourism*. Oxford: Butterworth Heinemann.
Hatton, M. (2004), "Current Issues Paper: Redefining the relationships—The future of travel agencies and the global agency contract in a changing distribution system", *Journal of Vacation Marketing*, 10(2), 101–108.
International Council of Cruise Lines (2004), The cruise industry 2003 economic survey: Business Research and Economic Advisors.
International Council of Cruise Lines (2005), ICCL gambling guidelines: Policy statement. Retrieved 22 March 2005, from http://www.iccl.org/policies/gambling.cfm
Knowles, T., Diamantis, D., and Bey El-Mourhabi, J. (2004), *The globalisation of tourism and hospitality* (2nd ed.). London: Thomson Learning.
Luxury Alliance (2005), The world's finest travel experience. Retrieved 22 March 2005, from http://www.luxury-alliance.com/
Mancini, M. (2000), *Cruising: A guide to the cruise line industry*. Albany NY: Delmar.
Princess Cruises (2003), Webpage, from http://www.princess.com/ships/ap/
PSARA (2005), Benefits of PSARA membership. Retrieved 22 March 2005, from http://www.psa-psara.org/application.html
Spartan Travel (2005), Cruising styles. Retrieved 22 March 2005, from http://spartan.travwell.net/cruises/choosing/style/
Vellas, F., and Becherel, L. (1995), *International tourism*. Basingstoke: Macmillan Press.
World's Leading Cruise Lines (2005), Anywhere you want to go, anyway you want to feel. Retrieved 22 March 2005, from http://www.worldsleadingcruiselines.com/intro.html

3

Maritime Issues and Legislation

Learning Objectives

By the end of the chapter the reader should be able to:

- Understand how the cruise industry interfaces with shipping in general
- Reflect on the commercial nature of shipping
- Consider the legal environment for the cruise industry
- Identify the role of international maritime organizations for cruise operations

In Chapter 1 the development of the cruise industry was described from a historical perspective. This chapter seeks to take a more holistic view of the contemporary nature of shipping so as to develop a greater depth of understanding about the cruise industry in context. There are many constraints for operators in the international shipping arena that are in place to ensure certain safeguards exist. Therefore, the legal environment is also examined to highlight critical factors. Legal issues are paramount for a number of organizations that have vested interests in either developing the legal framework or in supporting operators to comply appropriately. For this reason, an overview of the major maritime organizations is provided to illustrate their importance and involvement for the cruise industry and the maritime industry.

The Shipping Industry

The cruise industry is a type of passenger travel that arose phoenix-like after the Second World War, when jet planes were introduced as mass transportation vehicles to replace the stately and seemingly invincible transatlantic liners (Dickinson and Vladimir, 1997). Over the last three decades, the renaissance of cruising has been relentless and, for the large cruise corporations, it has also been highly lucrative. A number of issues yet remain that have a broader impact, from a shipping point of view, in terms of operational effectiveness, fair trading, environmentalism, and safety.

The shipping industry is, according to Farthing and Brownrigg (1997), the most international of all industries. This reflects the nature of transporting cargo or goods and people across seas and oceans internationally and the nature of the ships and their crew that are frequently multinational. However, the shipping industry is actually better described as a collection of industries (Farthing and Brownrigg, 1997) as shown in Table 3.1.

According to Lloyd's Register of World Shipping (2004) there were 89,899 ships weighing a combined 605,218,000 GRT in 2003. This puts into stark relief the 255 ships that Ward (2005) identifies as the world cruising fleet in 2004. The Institute of Shipping Economics and Logistics (ISL)

Table 3.1: **The components of shipping**

Wet bulk	Carry wet cargoes such as oil, chemicals, petroleum or anything in liquid form in tanks or specially designed holds (ships may be called tankers)
Dry bulk	Carry dry commodities such as iron ore, coal, grain, fertilisers, and sugar
Cargo liners	Scheduled vessels that carry containers or space onboard to a specific timetable
Coastal and short sea	Sometimes called tramp ships, these vessels offer alternative means of transporting goods rather than using road or rail
Cruise ships or passenger liners	Cruise ships are more common than passenger liners although some services such as Cunard still provide some liner services
Ferries	Tend to provide liner-like scheduled services with facilities to carry people, cars and other transportation
Offshore operations	This sector includes oil and gas rigs and supports exploration for mineral extraction at sea

comments that, in 2003, around 75% of the world cruise fleet was owned by three major corporations: Carnival Corporation, 41.7%; Royal Caribbean, 22.9%; and Star Cruises, 8.9% (ISL, 2003). Global ownership may be consolidating, but there is still evidence that a diverse range of ship management and ownership comes from outside this group of owner-managers (Panaydes, 2001), including those who charter, lease, and purchase management services.

Some companies, such as Louis Cruises, own a fleet of vessels, some of which are chartered to tour operators (Louis Cruises, 2005). Others, like V Ships, are involved in supplying crew and management services for cruise companies (V Ships, 2005). The *Hanseatic* is an example of a small vessel that is currently on long-term lease to German cruise operator Hapag Lloyd (Cruises, 2005). This complex pattern of ownership and management is fundamental for many operators involved in the contemporary cruise industry.

The Legal Environment

According to Farthing and Brownrigg (1997), the notion of freedom of the seas stems from principles that were set out in the United Nations (UN) Law of the Sea Convention in 1982, which came into force in November 1994. This convention created an umbrella approach for virtually all activities undertaken in, over, or under the sea (including actions on and below the seabed). An important component of the legislation was the recognition that states possess an Exclusive Economic Zone (EEZ) that extends 200 nautical miles seaward. This convention allows freedom of navigation, rights of access or passage to shipping, or both, on the high seas, with certain provisos concerning access to the EEZ. The regulation is an example of a collective international agreement that is established for the benefit of all signatories to the UN in order to allow for free enterprise, open competition, and economic freedom.

Ship Nationality, Registration, and Flag

The terms "nationality", "registration", and "flag" are sometimes used as if they were synonymous, but that is not totally accurate. Indeed, a ship may be deemed to have the nationality of a state even if there is no evidence of documentation for that nationality and the ship is unregistered. When a ship is registered, it is recorded officially and indicates that the ship possesses a certain nationality. The registration sets in place the framework for any legal consequences attributed to the ship's owner, the ship's managers, and the ship's crew. In public law, registration allocates the ship to a specific state, together with the jurisdiction that applies from that state, and protection from that state including the right to fly that state's flag. In private law the registration creates protection for the title of the owner and those who may hold securities in the form of financial interests of the vessel. The flag is symbolic and flown at the ship's stern as a mark of identification but otherwise the term "flag" is shorthand for the nationality of a vessel (Farthing and Brownrigg, 1997).

The implications of nationality for a ship and its owner present serious issues, apart from the aforementioned legal aspects, that can affect operational costs. Some countries require that ships registered in that country be crewed either entirely by nationals or a given percentage of nationals. For example the crewing, ship construction, and ownership requirements to flag a vessel in the United States are said to be among the most restrictive of the maritime nations. Current manning regulations for US-flag vessels engaged in coastwise trade mandate that all officers and pilots and 75% of other on board personnel be US citizens or residents. In addition, US-flag vessels engaged in coastwise trade must be owned by US citizens and constructed in US shipyards. This construction requirement applies to the entire hull and superstructure of the ship and the majority of all materials outfitting the vessel.

A cruise ship has many options for registration with states or countries that may be other than the owner's nationality. The reasons and benefits for this are many, including:

- Neutrality in the event of conflict
- Reduced tax liability
- Reduced registration fees
- Reduced crewing costs

Panama, Liberia, Cyprus, the Bahamas, and Malta were stated by Farthing & Brownrigg (1997) to be five of the world's largest fleets, suggesting that these states operated more liberal, economically attractive conditions and that they were seen to be effective and efficient in supporting the needs of ship operators. According to the ISL, nearly half of the world cruise fleet is now attributable to the Bahamas and Panama. The Bahamas, Panama, and Liberia had previously dominated the cruise shipping industry, but in 2003, nineteen vessels switched registration from Liberia to Panama because of the unstable political situation in the west African country (ISL, 2003).

According to the ICCL (2005), the major countries offering flags of registry for cruise vessels are the United Kingdom, Liberia, Panama, Norway, Netherlands, the Bahamas and, despite the stringent regulations, the United States. All of these countries are member states of the International Maritime Organization (IMO), an organization that is centrally important for maritime developments relating to safety.

The ICCL identifies a number of factors that must be met for a valid registry. One is that a flag state must be an IMO member nation, which has adopted all of the IMO's maritime safety resolutions and conventions. Secondly, a flag state should have an established maritime organization that is capable of enforcing all international and national regulations. Major flag registries are said to provide comprehensive maritime expertise and administrative services. In addition they are required to conduct annual safety inspections prior to the issuance of a passenger vessel certificate and use recognized classification societies to monitor its vessels' compliance with all international and flag state standards.

Marine Pollution

MARPOL, the International Convention for the Prevention of Pollution from Ships, is an abbreviation that is formed by the first three letters of "marine" and "pollution." The MARPOL agreement has been ratified by approximately 90 nations, including the United States and most other major maritime nations of the world. It governs a broad range of maritime issues relating to potential marine pollution, including oil, chemicals, garbage, and sewage, and mandates proper disposal or discharge. All ships operating in the United States must also comply with US regulations, including the Clean Water Act and the Oil Pollution Control Act. Likewise, ships operating in other countries must also pay due regard to additional regulations that may apply. In the United States, the cruise industry works with the US Coast Guard, the Environmental Protection Agency, and other federal and state regulators, as well as maritime groups, such as the Center for Marine Conservation and Ocean Advocates, to find productive environmental solutions.

Table 3.2: **ICCL industry standard E-01-01 Revision 2 (2003)**

Members of the International Council of Cruise Lines are dedicated to preserving the marine environment and in particular the pristine condition of the oceans upon which our vessels sail. The environmental standards that apply to our industry are stringent and comprehensive. Through the International Maritime Organization, the United States and other maritime nations have developed consistent and uniform international standards that apply to all vessels engaged in international commerce. These standards are set forth in the International Convention for the Prevention of Pollution from Ships (MARPOL). In addition, the United States has jurisdiction over vessels that operate in US waters where US laws, such as the Federal Water Pollution Control Act, the Act to Prevent Pollution from Ships, and the Resource Conservation and Recovery Act which applies to hazardous waste as it is landed ashore for disposal, apply to all cruise ships. The US Coast Guard enforces both international conventions and domestic laws. The cruise industry's commitment to protecting the environment is demonstrated by the comprehensive spectrum of waste management technologies and procedures employed on its vessels.

ICCL members are committed to:

- Designing, constructing and operating vessels so as to minimize their impact on the environment;
- Developing improved technologies to exceed current requirements for protection of the environment;
- Implementing a policy goal of zero discharge of MARPOL, Annex V solid waste products (garbage) by use of more comprehensive waste minimization procedures to significantly reduce shipboard generated waste;
- Expanding waste reduction strategies to include reuse and recycling to the maximum extent possible so as to land ashore even smaller quantities of waste products;
- Improving processes and procedures for collection and transfer of hazardous waste; and
- Strengthening comprehensive programs for monitoring and auditing of onboard environmental practices and procedures in accordance with the International Safety Management Code for the Safe Operation of Ships and for Pollution Prevention (ISM Code).

MARPOL is interpreted by the ICCL (ICCL, 2005) to make waste management by the cruise industry operational (Table 3.2). In respect of industry waste management standards, the cruise operators who are members of ICCL have agreed to incorporate the following standards for waste stream management into their safety management systems.

"Graywater" and "blackwater" are types of wastewater produced by ships carrying passengers or crew. Graywater is produced by showers, sinks, or basins and in food preparation, while blackwater refers to sewage. On cruise ships, both are treated in accordance with industry regulatory requirements that are frequently more stringent and demanding than government regulations (Table 3.3).

In the United States, the Coast Guard enforces regulations regarding ocean dumping from vessels. Under US regulations it is illegal to dump plastic refuse and garbage mixed with plastic into any waters. In addition, the regulations restrict dumping of non-plastic trash and other forms of garbage. These regulations apply to all US vessels wherever they operate (except in waters under exclusive jurisdiction of another state) and to foreign vessels operating in US waters out to and including the Exclusive Economic Zone (200 miles offshore).

Safety of Life at Sea

The International Convention for the Safety of Life at Sea (SOLAS) was first adopted in 1948. It is called a "living" document—that is, one that is continuously amended and updated. SOLAS is concerned with the establishment of international regulations that address maritime safety, including lifesaving, fire protection, and ship stability. According to the US Coast Guard, cruise ships are regulated for safety by government agencies in the following way (US Coast Guard, 2004). See Table 3.4 for more information.

While some vessels are registered in the United States, current patterns suggest that most are not, and for these vessels the safety inspection is administered within the country of registration. The US Coast Guard requires any ship, irrespective of country of registration, to meet the SOLAS

Table 3.3: **Waste management standards (2003)**

1. Photo Processing: Including X-Ray Development Fluid Waste: Member lines have agreed to minimize the discharge of silver into the marine environment through the use of best available technology that will reduce the silver content of the waste stream below levels specified by prevailing regulations.
2. Dry-Cleaning Waste Fluids and Contaminated Materials: Member lines have agreed to prevent the discharge of chlorinated dry-cleaning fluids, sludge, contaminated filter materials and other dry-cleaning waste by-products into the environment
3. Print Shop Waste Fluids: Member lines have agreed to prevent the discharge of hazardous wastes from printing materials (inks) and cleaning chemicals into the environment.
4. Photo Copying and Laser Printer Cartridges: Member lines have agreed to initiate procedures so as to maximize the return of photo copying and laser printer cartridges for recycling. In any event, these cartridges will be landed ashore.
5. Unused and Outdated Pharmaceuticals: Member lines have agreed to ensure that unused and/or outdated pharmaceuticals are effectively and safely disposed of in accordance with legal and environmental requirements.
6. Fluorescent and Mercury Vapor Lamp Bulbs: Member lines have agreed to prevent the release of mercury into the environment from spent fluorescent and mercury vapor lamps by assuring proper recycling or by using other acceptable means of disposal.
7. Batteries: Member lines have agreed to prevent the discharge of spent batteries into the marine environment.
8. Bilge and Oily Water Residues: Member lines have agreed to meet or exceed the international requirements for removing oil from bilge and wastewater prior to discharge.
9. Glass, Cardboard, Aluminum and Steel Cans: Member lines have agreed to eliminate, to the maximum extent possible, the disposal of MARPOL Annex V wastes into the marine environment. This will be achieved through improved reuse and recycling opportunities. They have further agreed that no waste will be discharged into the marine environment unless it has been properly processed and can be discharged in accordance with MARPOL and other prevailing requirements.
10. Incinerator Ash: Member lines have agreed to reduce the production of incinerator ash by minimizing the generation of waste and maximizing recycling opportunities.
11. Graywater: Member lines have agreed that graywater will be discharged only while the ship is under way and proceeding at a speed of not less than 6 knots; that graywater will not be discharged in port and will not be discharged within 4 nautical miles from shore or such other distance as agreed to with authorities having jurisdiction or provided for by local law except in an emergency, or where geographically limited. Member lines have further agreed that the discharge of graywater will comply with all applicable laws and regulations.
12. Blackwater: ICCL members have agreed that all blackwater will be processed through a Marine Sanitation Device (MSD), certified in accordance with US or international regulations, prior to discharge. Discharge will take place only when the ship is more than 4 miles from shore and when the ship is traveling at a speed of not less than 6 knots.

convention if they wish to take on vessels in US ports. US law expects that any cruise company advertising in the United States will disclose the country of registration for their vessels. SOLAS is far reaching in its remit and requires compliance with stringent regulations regarding structural fire protection, firefighting and lifesaving equipment, watercraft integrity and stability, vessel control, navigation safety, crewing and crew competency, safety management, and environmental protection.

The Coast Guard, in respect of SOLAS requirements, examines all cruise ships when they first visit US ports. Thereafter, the vessels are inspected, or checked for compliance, quarterly. Records relating

Table 3.4: **Safety oversight undertaken by the US Coast Guard (US Coast Guard, 2004)**

In terms of vessel safety, cruise ships of US registry must meet a comprehensive set of Coast Guard safety regulations and be inspected annually by the Coast Guard to check for compliance. The safety regulations include such things as hull structure, watertight integrity, structural requirements to minimize fire hazards, equipment requirements for lifesaving, fire-fighting, and vessel control, and requirements pertaining to the safe navigation of the ship. If the ship passes its annual inspection, it is issued a Coast Guard Certificate of Inspection valid for one year. The certificate must be displayed where passengers can see it.

to these inspections (called Control Verification Examinations) are available for public scrutiny. Inspectors involved with these examinations board the ship to verify structural fire safety, ensure that lifesaving equipment is available and located as required in the appropriate condition, witness fire and abandon ship drills as conducted by the ship's crew, and test key equipment such as steering systems, fire pumps, and lifeboats. The Coast Guard has the authority to require correction of any deficiencies before allowing the ship to take on passengers at the US port.

In terms of crew member competence, the Coast Guard can suspend or revoke licenses or merchant mariner's documents if a US-registered ship is found to be operating below published standards for experience and training. On foreign-flag ships, SOLAS requires that the ships must be efficiently and sufficiently staffed, and this is checked during control verification examinations. SOLAS is not designed to provide guarantees for health care, and, as a result, it is not a proviso that cruise ships carry a ship's doctor.

SOLAS requires that the ship's captain schedule and implement periodic fire and lifeboat drills. This is intended both to give the crew practice and to show passengers the critical action that may be required in the event of a serious incident or emergency on board. For this reason SOLAS expects that all passengers participate in these drills. The drills are scheduled according to the duration of the cruise. In a one-week cruise, the first drill would take place as soon as all passengers were on board and immediately prior to sailing. If the cruise lasts more than a week, an additional drill would take place every week thereafter. For a cruise lasting less than a week, the drill takes place within 24 hours of departure from the home port.

Notices are to be posted in clear view in every passenger cabin or stateroom to provide easily understood information regarding safety issues. This notice includes:

- How to recognize the ship's emergency signals (alarm bells and whistle signals are normally supplemented by announcements made over the ship's public address system).
- The location of passenger life preservers in that stateroom (special life preservers will be provided for children, if necessary, by the room steward).
- Instructions and pictures explaining how to put on the life preserver, as well as the lifeboat assignment for passengers in specific staterooms. Modern cruise ships carry a variety of survival craft. Passengers are invariably assigned to lifeboats or similar survival craft that can be used in an emergency.

Crew members from the hotel department play an important and potentially critical part in the safety routines and are generally responsible for assisting and directing passengers for emergency drills, although some may have other safety duties. The regulations call for direction signs showing the path to reach lifeboats to be posted in passageways and stairways throughout the ship. The crew member in charge of each lifeboat will gather or muster the passengers assigned to that lifeboat and give passengers any final instructions for properly donning and adjusting their life preservers. The crew should be prepared to help passengers and clarify the emergency procedures, if necessary.

Sanitation and Cleanliness

In the United States, the responsibility for maintaining oversight of sanitary conditions on passenger vessels is undertaken by the Public Health Service (USPHS). USPHS conducts both scheduled and unscheduled inspections of passenger vessels in US ports under its Vessel Sanitation Program (VSP), focusing on proper sanitation for drinking water, food storage, food preparation and handling, and general cleanliness. USPHS provides the public with results of inspections on individual vessels and takes reports of unsanitary conditions on individual vessels. In other countries, similar inspections are undertaken by state bodies. For example, the Australian Quarantine Inspection Service, the UK Port Health Authority, and the Canadian Public Health Bureau appoint environmental health officers to implement such inspections.

Cruise companies take these inspections very seriously because it is in the best interest of the cruise company to comply, to be safe, and to secure high scores. More details about this process can be found in Chapter 10.

Marine Security (MARSEC)

MARSEC has been developed to establish regulations for crew competence that apply to training, certification, and watch-keeping, so as to ensure safe practice and secure environments for passengers and crew. MARSEC was introduced specifically to respond to potential risks following the terrorist attacks of September 11, 2001. Prior to MARSEC, the International Maritime Organization monitored the International Safety Management code (ISM), which covered both mandatory safety and antipollution standards. MARSEC includes published Standards for Training, Certification, and Watch-keeping (STCW). In July 2004 a code of practice was introduced internationally to address heightened tensions about the safety of shipping. The International Ship and Port Facility Security (ISPS) code is examined in Chapter 10, but the following information provides a summary of key elements.

MARSEC requires that ships carry a designated vessel security officer and that this person be responsible for the ship's security plan. The security officer is expected to be a senior deck officer who has responsibility for standing watch. The responsibilities include developing the ship's security plan, ensuring that appropriate adequate training is provided for officers and crew, ensuring that the ship complies with the security plan, and maintaining knowledge relating to international laws, domestic regulations, current security threats, and patterns relating to security issues.

The ship's security officer acts as a liaison between the ship, relevant authorities, and the company's security officer. Typically, this individual would be involved with undertaking risk assessments, developing strategies, and evaluating points of vulnerability. MARSEC operates three levels of security status:

- Level 1—minimum appropriate security measures required
- Level 2—heightened risk of security incident
- Level 3—probable or imminent security incident for a limited time

Threat levels are communicated by ports to ships in timely fashion, so the ship can have sufficient time to consider best action. The ship's master may elect to elevate a threat level if the threat is considered above that stated by the port.

In the United States, according to federal regulations, terminal operators and cruise lines share the primary responsibility for shoreside and shipboard security of passengers. The Coast Guard examines all security plans and can require improvements in their security measures. Passengers embarking on international voyages may expect to have their baggage searched or passed through screening devices before boarding. The terminal operator and cruise line have strict procedures for passenger identification and visitor control. Passengers who wish to have friends visit the ship prior to sailing should check with the cruise line well in advance. All these security measures are designed to prevent the introduction of unauthorized weapons and persons on the cruise ship. More details on this subject are included in Chapter 10.

Financial Responsibility

The US Federal Maritime Commission requires that operators of passenger vessels carrying 50 or more passengers from a US port must be financially secure and capable of reimbursing their customers if the cruise is cancelled. The Commission also requires proof of ability to pay claims arising out of passenger injuries or death for which the ship operator may bear some liability. The Commission does not have the legal authority to automatically secure these financial settlements for individual consumers.

If a cruise is cancelled or there is an injury incurred during the cruise, the consumer will have to initiate action on his or her own behalf against the cruise line. Insurance for shipping is provided by many of the world's largest financial and insurance companies, such as Lloyd's of London, Lloyd's of America, and the American Institute of Marine Underwriters.

Maritime Organizations

It is important to recognize certain organizations involved in the maritime industry or the cruise industry. A number of these organizations, some of which have been previously mentioned, are listed and described below.

IMO (International Maritime Organization)

The International Maritime Organization (formerly known as the Intergovernmental Maritime Consultative Organization) was established in 1948 as an agency of the United Nations to set international maritime policy and regulate the shipping industry. In this capacity, it develops a cross-governmental, consensual approach for safety and practices at sea. The IMO is the glue that binds together the treaties and conventions for international shipping, with responsibility for ensuring compliance with implementation of regulations, although the principal responsibility for enforcing these regulations rests with the flag states, or the country within which the ship is registered. "Port state control" supplements flag-state enforcement by allowing officials from any country a ship may visit to inspect foreign flag ships to ensure that they comply with international requirements.

The IMO's slogan, "Safe, secure, and efficient shipping on clean oceans," encapsulates the agency's mission statement. Despite this seemingly mammoth task, the organization remains relatively lean in scale because of the requirement for individual countries to undertake enforcement. The US Coast Guard represents the United States in this international agency. The IMO has been instrumental in the development and adoption of several important treaties or conventions, including the previously mentioned SOLAS agreement, the International Convention for the Prevention of Pollution from Ships (known as the MARPOL agreement), and the International Safety Management Code (ISM), which is part of SOLAS and the Standards for Training, Certification, and Watch-keeping (STCW).

Classification Societies

These classification societies are mainly organizations whose primary function is to inspect ships at regular intervals to ensure they are seaworthy and regularly maintained in keeping with the classification societies' rules. Classification societies also inspect cruise ships for compliance with international safety regulations, including SOLAS, STCW, and MARPOL. Major classification societies include the American Bureau of Shipping in the United States, Lloyd's Register of Shipping in the United Kingdom, Det Norske Veritas in Norway, Bureau Veritas in France, and Registro Italiano Navale Group in Italy.

Cruise Lines International Association (CLIA)

The CLIA is a marketing and promotion organization that represents 23 member cruise lines and approximately 19,000 North American travel agencies. The CLIA was formed in 1975 with the specific intent of promoting the benefits of cruising. The CLIA also undertakes training in line with its mission "to educate travel agents and to promote the value, desirability, and affordability of the cruise vacation experience." The CLIA joined with the International Council of Cruise Lines to establish the Cruise Line Coalition in 2001 to act as an information source for the industry.

Table 3.5: **ICCL statements of purpose (International Council of Cruise Lines, 2004c)**

That the cruise industry:

- provides a safe, healthy, secure and caring ship board environment for both passengers and crew,
- minimizes the environmental impact of its vessel operations on the ocean and marine life,
- adheres to regulatory initiatives and in leading the effort to improve maritime policies and procedures,
- creates a regulatory environment that will foster the continued growth of the industry,
- delivers a reliable, affordable and enjoyable cruise experience to people from all walks of life.

International Council of Cruise Lines (ICCL)

The ICCL is a trade association that represents a range of major cruise companies and associate members such as suppliers and industry partners. With its mission to "participate in the regulatory and policy development process and ensure that all measures adopted provide for a safe, secure, and healthy cruise ship environment," the association has an important role to play. To achieve its aims, the association analyzes and interprets international shipping policy and offers recommendations to its membership on a wide variety of issues, including safety, public health, environmental responsibility, security, medical facilities, passenger protection, and legislative activities. In undertaking its role, the ICCL works closely with key domestic and international regulatory organizations, policymakers, and other industry partners and serves as a nongovernmental consultative organization to the IMO.

In its capacity as a representative association, the ICCL aims to ensure that the objectives listed in Table 3.5 are achieved.

Florida-Caribbean Cruise Association (FCCA)

This is a trade organization that was inaugurated in 1972 to provide a forum for 13 cruise brands to meet and debate operational issues concerning them. The FCCA can highlight legislation, tourism development, port safety, security, and other emerging issues to create solutions that are a product of cooperation and partnership. The FCCA also undertakes targeted training, such as customer service programs for taxi drivers in ports, as well as commissioning research looking at the impacts of cruising. The association has created a charitable foundation to find humaritarian causes and help to improve the laws of people most in need.

North West Cruise Ship Association (NWCA)

The North West Cruise Ship Association is a not-for-profit body that represents nine cruise lines operating in Hawaii, Canada, Alaska, and the Pacific Northwest. The association was established in 1986 initially to focus on security concerns, although later it developed a broader role addressing government relations with respect to legal and regulatory issues. In addition, the association seeks to maintain positive links with the local communities involved in cruising areas in order to work on environmental protection, economic development, and other industry-connected concerns.

Summary and Conclusion

This chapter has examined a range of maritime issues that concern cruise ships and shipping in general. It has highlighted the importance of the IMO and commented on the range of regulations that are in place to help make shipping safe for crew and passengers. Various groups support the industry. Some, like the CLIA, have a role in marketing support, and others, such as the FCCA and the NWCA, have a geographical focus and represent cruise company interests in specific areas. The ICCL takes on a more political stance in working with governments and the IMO.

Glossary

Convention: An agreement and statement of purpose between members of a representative group.
Neutrality: A position where the individual doesn't take sides in a conflict.
Regulation: A rule that is stated and must be complied with.
Superstructure: The part of a ship above the waterline.

Chapter Review Questions

Many of the following questions contain an industry acronym. In each case you should be able to state what each acronym stands for before answering the question.

1. What is the IMO and what does it do?
2. Explain the following:
   a. SOLAS
   b. MARPOL
   c. MARSEC
3. What is the purpose of the Vessel Sanitation Program?
4. Compare the ICCL, CLIA, FCCA, and NWCA.
5. What is "Port State Control"?

Additional Reading and Sources of Further Information

http://www.imo.org/index.htm: IMO
http://www.iccl.org/: ICCL
http://www.cruising.org/: CLIA
http://www.f-cca.com/: FCCA
http://www.nwcruiseship.org/: NWCA

References

Cruises (2005), Hapag Lloyd Hanseatic cruise ship overview. Retrieved 22 March 2005, from http://cruises.about.com/od/cruiseshipprofiles/ss/hanseatic.htm

Dickinson, R., and Vladimir, A. (1997), *Selling the sea: An inside look at the cruise industry*. New York: Wiley.

Farthing, B., and Brownrigg, M. (1997), *International shipping* (3rd ed.). London: LLP Ltd.

ICCL (2005), International maritime industry—background and facts. Retrieved 22 March 2005, from http://www.iccl.org/faq/imi.cfm

International Council of Cruise Lines (2004), What is the ICCL? Retrieved 8 April 2005, from http://www.iccl.org/whoweare/index.cfm

ISL (2003), Executive summary—SSMR market analysis no. 6. Retrieved 22 March 2005, from http://www.isl.org/productsservices/publications/samples/cruise.shtml.en

Lloyd's register (2004), *World Fleet Statistics* 2003. London: Lloyds.

Louis Cruises (2005), Charters. Retrieved 22 March 2005, from http://www.louiscruises.com/

Panaydes, P. M. (2001), *Professional ship management*. Aldershot: Ashgate Publishing.

US Coast Guard (2004), Cruise ship fact sheet. Retrieved 22 March 2005, from http://www.uscg.mil/hq/g-m/factsheetcruiseship.doc

V Ships (2005), Leisure management. Retrieved 22 March 2005, from http://www.vships.com/

Ward, D. (2005), *Complete guide to cruising and cruise ships 2004*. London: Berlitz Publishing.

4

Cruise Geography

Learning Objectives

By the end of the chapter the reader should be able to:

- Consider geography from a cruise industry perspective
- Evaluate the primary and secondary cruise sectors
- Identify major cruise ports in each sector
- Consider the attractions and features that are important in defining a cruise port and destination

In a practical sense, cruise companies regard the world as a series of sectors that meet various market needs. For the largest brands, this outlook allows companies to configure operations to take account of:

- Seasonality, weather patterns, and optimum conditions for cruising
- Sales and marketing
- Supply and servicing of ships

This chapter considers the influence and effect of geography on the cruise industry. To start with, it is impossible to consider cruising without reflecting on the conditions that arise from the prevailing climate. Passenger comfort and safety are directly affected if a cruise ship sails in a particular part of an ocean or sea at a particular time of the year. This also holds true for destinations visited and shore activities that may be offered.

To avoid potential discomfort for customers, cruise ships tend to steer clear of parts of the world where difficult sea conditions occur because of geography, climate, and seasonal variations (Burton, 1995). Many stories are told of severe weather conditions in specific locations. For example, the Bay of Biscay, the Cape of Good Hope, the Bay of Bengal, and the North Atlantic have reputations suggesting they can provide extremes of weather for seafarers or navigators. Yet knowledge of weather patterns and records of tidal variations permit cruise operators to predict where ships can travel with a high degree of safety to enable virtually all the world's oceans and seas to be traversed and all coastal ports to be visited (see Figure 4.1).

Weather Cycles

Weather patterns are complex. They are influenced by many factors, including the sun's rays, the world's rotational axis (which tilts 23.5° from the perpendicular, thus creating seasonal variations), the land masses and oceans, currents, and the moon's gravitational pull (which creates tidal variations).

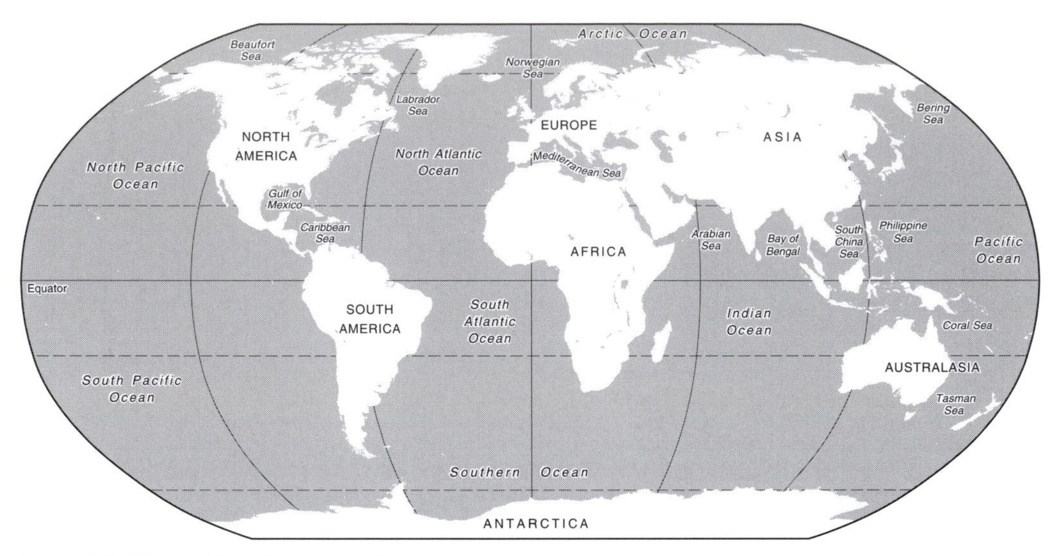

Figure 4.1: The world: land masses and oceans

Northern and Southern hemispheres experience seasons at opposite times of the year, reflecting the position of the world as it orbits the sun (Rees, 1992).

Tropical Zones

The point where the Earth is closest to the Sun is known as the equator. The Tropics of Cancer and Capricorn are lines of latitude that run parallel to the equator at 25° North and 25° South, respectively. These two boundaries define the region that is known as the "tropics." Points above and below the equator can be affected by bad weather and storms, although the equator can be calm. The weather effect when the wind and sea are calm is known as the doldrums.

Tropical cyclones are triggered by latent heat, water condensation, and cloud formations. These can be monitored and, to a degree, patterns can be predicted so that ships are forewarned and can take appropriate measures. Most contemporary cruise ships, with the odd exception, are designed for cruising in relatively benign conditions and, therefore, itineraries are influenced in part by weather patterns. Cyclones can create winds in excess of 120 km (75 miles) per hour. The specific names of tropical cyclones depend on the area (Rees, 1992):

Atlantic and Eastern Pacific—hurricanes
Western Pacific—typhoons
Philippines—baguios
Australia—willy-willy
Indian Oceans—cyclones

Tourists and Climate

Cruise ships tend to focus on warm, temperate climates and calm seas. However, continuous growth in sectors such as Alaska, the emergence of Antarctica and the southernmost areas of South America interest in northerly sectors such as Iceland, Scandinavia, and the Baltic ports are testament to the diversity of choice that is now available for cruise tourists.

Table 4.1: **Temperature and clothing zones**

Latitude	*Temperature zone and climatic type*	*Corresponding clothing zone*
Equator	Hot—equatorial, tropical and dessert	Minimum clothing and light protective clothing
	Warm temperate—Mediterranean and eastern margin climates	One layer clothing
	Cool temperate—marginal and continental types	Two layer clothing
	Cold climates	Three layer clothing
Poles	Arctic and polar climates	Maximum clothing

Invariably, tourists make judgments about what parts of the world to visit by taking into account a broad spectrum of personal circumstances and by accessing new information or relying on prior learning about the place to be visited (Bansal and Eiselt, 2004; Gibson, 2004). These decisions are highly personalized and are likely to include a desire to learn new things, a drive to satisfy personal motives, the need to address latent curiosity, the opportunity to relax and escape routines, and the requirement to experience a different climate (Bansal and Eiselt, 2004).

This need to identify an appropriate and desirable environment for a vacation raises questions about which climate tourists regard as comfortable. According to Burton (1995), it is possible to identify comfort zones for various tourist activities that take into account factors such as temperature, humidity, wind, rainfall, clouds, and sunshine. Based on an analysis of world climates, Burton presents a five-stage model to describe tourists' clothing regimes in relation to climate type (Table 4.1).

Primary Cruising Regions: The Caribbean

The Caribbean currently attracts more passengers than any other region in the world. According to Wild and Dearing (2005), recent growth patterns in the Caribbean emerged because North American passengers were seeking cruises that offered certain characteristics, and one of these was the need to remain close to home. The three years since 9/11 consolidated the Caribbean's position as the number one cruise region, and while growth may have slowed since then, the position still reflects an upward trend.

Calculating passenger figures that affect cruise destinations can be accomplished in a number of ways. For example, potential passenger throughput (PPT) takes the number of cruises and multiplies the total number of passengers for each vessel to create a quick reference number; the passenger per night (PN) calculates total numbers of nights spent in individual ports. As would be expected in terms of scale of operations, vessels operated by brands owned by Carnival Corporation, the world's largest cruise company, dominate in this region, with 54% of all passenger nights (PN) compared with 33% for Royal Caribbean (Wild and Dearing, 2005). According to the data produced by Wild and Dearing, the western Caribbean is expected to attract larger numbers of passengers during 2005 (estimated at 2.75 million) than the eastern and southern Caribbean (an estimated 2.2 million).

Burton (1995) describes the Caribbean (see Figure 4.2) as a "4000 km arc which sweeps eastward from Florida to the Venezuelan coast." The islands are diverse in physical character and climate, accessibility, historical background, and political setting. According to industry sources and despite the obvious attractions of warm, crystal-blue seas and palm-tree-ringed beaches for relaxation and swimming, another primary activity for many customers visiting some Caribbean islands is shopping. Ports such as Nassau have been tagged as a duty free paradises.

The layout of the islands presents opportunities for cruise companies to create varied itineraries that incorporate a number of contrasting islands. These itineraries can start with embarkation at a port in Florida such as Fort Lauderdale, Port Everglades, Port Canaveral, or Miami. Alternatively, vessels can create an itinerary originating in a Caribbean island such as Puerto Rico or Barbados. In recent years

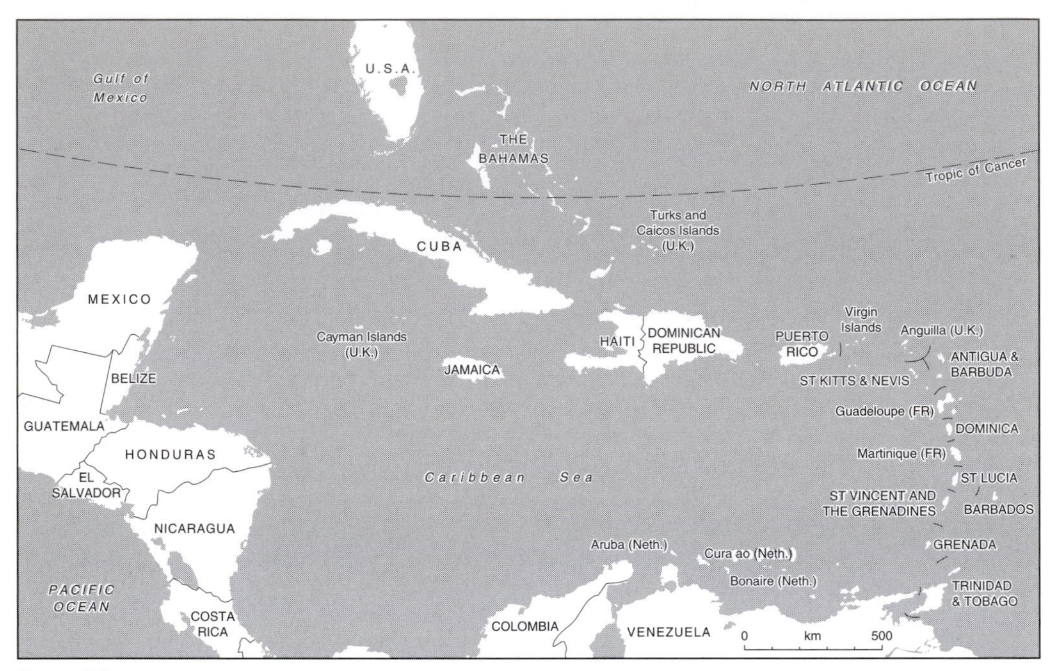

Figure 4.2: The Caribbean

southern and eastern Caribbean islands have fared less well than western Caribbean islands, because itineraries have switched from 2-week tours to shorter 4- or 7-day excursions (Burton, 1995). Competition for the tourist dollar has led to a situation where port fees for many Caribbean islands are relatively inexpensive, at between US$4 to US$6 per customer. In this way, a vessel such as the Diamond Princess, with 2,500 customers, would pay approximately US$12,500 in port fees. In addition, cruise companies own some islands: Royal Caribbean has ownership of Coco Cay, Holland America owns Half Moon Cay, Disney Cruises owns Castaway Cay, Norwegian Cruise Lines owns Great Stirrup Cay, and Princess Cruises has an island called Princess Cay. The benefits of such ownership include generating revenue from shore-based activities and controlling costs associated with ports of call (a Cay, or Key as it is known in the United States, is a low island or reef of sand or coral).

Some islands are less tranquil or accommodating to tourists than may be expected. Cuba, the largest of the Caribbean islands and located 145 km south of Florida, has a history of being opposed to the politics of the United States and this posture has inhibited US tourist trade growth. Other recent examples include political instability in Haiti (BBC News, 2004) that has had a detrimental effect on the island's economy and has negatively affected cruise visits. However, in general, the situation on the majority of these "island paradises" is calm and settled.

Major Island Destinations

The Cruise Line Industry Association (CLIA, 2005a) describes three specific parts to the Caribbean—the eastern Caribbean and the Bahamas, the western Caribbean and the southern Caribbean. Selected destinations from these areas are described in this section. The Caribbean Tourism Organization (2005) is a trade organization that represents many of the Caribbean islands: Anguilla, Antigua and Barbuda, Aruba, Bahamas, Barbados, Belize, Bermuda, Bonaire, the British Virgin Islands, the Cayman Islands, Cuba, Curacao, Dominica, Grenada, Guadeloupe/St. Barts/St. Martin, Guyana, Haiti, Jamaica,

Martinique, Montserrat, Puerto Rico, St. Eustatius, St. Kitts and Nevis, St. Lucia, St. Maarten, St. Vincent and the Grenadines, Suriname, Trinidad and Tobago, the Turks and Caicos Islands, and the US Virgin Islands. Some of these islands attract large volumes of cruise passengers. For example, Aruba, with a population of 75,000 (CIA, 2005), attracted 228 cruise ships carrying 410,962 passengers in the cruising season of October 2003 to April 2004 (Aruba Cruise Tourism, 2005).

Eastern Caribbean: The Bahamas

The islands of the Bahamas are close to the Caribbean but are not part of this region (Mancini, 2000). However, the Bahamas' close proximity to both the South Florida Coast and the eastern Caribbean makes them a natural itinerary option for cruise planners, which means the islands are a frequent stop in Caribbean cruises. The combined Bahamas and eastern Caribbean area is relatively accessible from US ports such as Miami, Port Everglades, and Port Canaveral, as well as from San Juan in Puerto Rico. However, because of the cumulative distance involved for this type of itinerary, the duration for some cruises is likely to be in excess of 7 days. There is a diversity of ports in the region, including the aforementioned cays and islands that are privately owned by cruise companies (Caribbean Tourism Organization, 2005). Several ports are described below, followed by a table that presents information about population, language, and currency. In all cases, for this and subsequent tables relating to destinations that appear in this chapter, populations of the port or island community are approximate.

The Bahamas

Nassau and Freeport on New Providence Island are the primary ports of call in the Bahamas. In 2004, Nassau was the sixth most visited port in the world and Freeport was the sixteenth (Wild and Dearing, 2004a). The name Bahamas is a derivation of the Spanish "Baha Mar," or shallow sea, and there are approximately 700 islands in this popular self-styled "paradise archipelago" of sun, sea, and sand. The beaches are held in high regard, but the islands also offer a variety of attractions beyond miles of white or pink sand. The islands claim the world's third largest barrier reef and a diversity of sea life, including whales and dolphins. The Bahamas have a population of 302,000 (70% on New Providence Island) and relys on tourism for 50% of employment and gross domestic product (GDP), or the total amount of revenue generated from sales of products and services. The islands have historical ties with the UK, as evidenced by cars that still drive on the left, despite many of the cars being manufactured as left-hand drive (Bahamas Tourism Office, 2005). Shopping, golf, and gambling are all available for tourists to the islands (Dervaes, 2003). The Bahamas were awarded the accolade of the Caribbean's leading destination in 2004 (World Travel Awards, 2004). The World Travel Awards are presented at an annual event. Travel agents from 200 different countries vote for the awards.

Puerto Rico

San Juan, Puerto Rico is both a port of call (or destination) and a base port. This dual role makes the island the seventh most visited destination in the world, according to Wild and Dearing (2004a). Puerto Rico is described as an "Island of Enchantment," with a broad range of multifaceted attractions including the archetypal tropical beach scene, diverse natural attractions, and a rich cultural heritage. Islanders came from a mix of cultures reflecting the scope of the island's origins, which includes African, Spanish, indigenous, and US influences. The population of Puerto Rico is just under 4 million. The currency is US dollars, and both English and Spanish are spoken (Puerto Rico Tourist Office, 2005).

St. Thomas, US Virgin Islands

St. Thomas and the island's port, Charlotte Amalie, are celebrated by shoppers. Over the years, the island has become a leading tax-free haven and this, combined with the natural allure of the scenery and the island attractions, creates a powerful draw (US Virgin Islands Tourism Authority, 2005). As a result, the island is the eighth most frequented port in the world (Wild and Dearing, 2004a). Cruise visitors to the islands have easy access to a shopping mall next to the pier and can also enjoy water sports, such as snorkeling and scuba diving expeditions, and land based activities (Dervaes, 2003). The term *scuba* is an acronym that stands for "self-contained underwater breathing apparatus."

St. Maarten

Philipsburg is St. Maarten's port. With one half Dutch and the other half French (referred to as St. Martin) the island has two national identities and two personalities. The half of the island where most cruise ships call at Philipsburg is Dutch. The island is the ninth most visited port in the world according to Wild and Dearing's (2004a) survey. Visitors enjoy beach activities, water based excursions, and cultural experiences when visiting this island (Mancini, 2000).

Antigua

This island is the eighteenth most visited port in the world by cruise passengers (Wild and Dearing, 2004a). Antigua is a verdant tropical island that boasts the historical attraction of Nelson's Dockyard, the eighteenth-century base for the British naval fleet (Mancini, 2000). The island is popular for snorkeling and scuba diving, and is said to be one of the sunniest of the eastern Caribbean islands (Antigua Barbuda Tourist Information, 2005). English is the first language for the island.

Other eastern Caribbean ports that are popular cruise destinations are Tortola, Dominica, St. Lucia, Martinique and St. Kitts.

Table 4.2: **Eastern Caribbean destination facts**

Destination	*Country*	*Region*	*Currency*	*Language*	*Population*
Nassau	Bahamas	East Caribbean	Bahamian Dollar	English	302,000
San Juan	Puerto Rico	East Caribbean	US Dollar	English/Spanish	3,916,000
St. Thomas	US Virgin Islands	East Caribbean	US Dollar	English	108,000
St. Maarten or St. Martin	Dutch/French	East Caribbean	Euro	Dutch/English and French/English	73,000
Antigua	Antigua	East Caribbean	East Caribbean Dollar	English	68,000

Western Caribbean

The western Caribbean is convenient for cruises that depart from Florida or ports such as Houston, Galveston, and New Orleans (Mancini, 2000). In addition, the itineraries for this region can be supplemented with Mexican destinations such as Cozumel (the third most visited port in the world according to Wild and Dearing (2004a), Cancun, Veracruz, and Tampico to create a varied and distinctive cruise program.

Key West

Key West is the southernmost point in the United States. Known as the "Conch Republic," Key West is famous for being the favorite haunt of artists, celebrities, presidents, and literary heroes such as Ernest Hemingway (Florida Keys and Key West Tourism Association, 2005). Yet the Florida Keys and Key West only became a fixture for visitors after an economic and social revival in the 1980s. The destination appears as the tenth most visited port in the world (Wild and Dearing, 2004a). The Keys' literary reference points, such as the homes of Ernest Hemingway and Tennessee Williams and former president Harry Truman's "Little White House," are high on the list of attractions; passengers may also go shopping or even go deep sea fishing in the Gulf of Mexico.

Cayman Islands

George Town in Grand Cayman is the main port of call and is the fifth most visited port in the world (Wild and Dearing, 2004a). The islands are famous for the opportunity to swim with stingrays, although many other attractions and experiences are available. The islands have a reputation for spectacular diving around the coral reefs, which are generously endowed with marine life. Grand Cayman Island is also home to the world's first sea turtle farm, the spectacular limestone and coral formations known as Hell, and the popular Seven Mile Beach (Cayman Islands Department of Tourism, 2005).

Jamaica

Jamaica is the Caribbean's second largest island, and Ocho Rios, Jamaica's port, is fifteenth in Wild and Dearing's survey (Wild and Dearing, 2004a). Jamaica has an array of natural wonders such Dunn's River Falls. Cruise passengers have the opportunity to climb the waterfall, take an expedition to the Blue Mountains, and go on an undersea tour or visits to caves (Visit Jamaica, 2005). The options are very wide, reflecting the natural and cultural diversity of the island. Music, as epitomized by the late Bob Marley, plays a part in Jamaican culture. Jamaica is the home of reggae and boasts a rich historical heritage.

Table 4.3: **Western Caribbean destination facts**

Destination	*Country*	*Region*	*Currency*	*Language*	*Population*
Key West	United States	Western Caribbean	US Dollar	English	24,800
Cayman Islands	Cayman Islands	Western Caribbean	Caymanian Dollar	English	44,200
Kingston	Jamaica	Western Caribbean	Jamaican Dollar	English	2,731,800

Southern Caribbean

This part of the Caribbean tends to be seen as more exotic because the islands are located close to Venezuela in South America and the itinerary usually requires using a home port from within the area, such as Barbados and Aruba. Many cruises to the southern Caribbean originate from San Juan in the eastern Caribbean and include a mixed itinerary of ports from both the eastern and southern Caribbean. This region enjoys the Caribbean's sunniest climate.

Barbados

Georgetown is the port for the island of Barbados which lies at the eastern edge of the southern Caribbean. The countryside has a softly rolling landscape, in contrast to some of the other volcanic islands that have been considered so far. Barbados has a strong British connection (Barbados Tourism Authority, 2005) and is a former colony (it gained full independence in 1966). Attractions include rum factory tours, touring the island, water sports, and visiting the many beautiful beaches.

Curaçao

Willemstad is Curaçao's capital. Curaçao is the main island of the group of islands known as the Dutch Antilles. Curaçao has an unmistakable Dutch heritage, reflected in its style of architecture (Curaçao Tourist Board, 2005). The island has a host of activities for cruise passengers, who may wish to visit the island's shops, the underwater park and Seaquarium, or the island's ostrich farm.

There are many other islands in this area including Bonaire, Trinidad, and Tobago. Itineraries may also include Venezuelan ports such as La Guaira (for Caracas or Venezuela) and Cartagena.

Table 4.4: **Southern Caribbean destination facts**

Destination	*Country*	*Region*	*Currency*	*Language*	*Population*
Barbados	United States	Southern Caribbean	Barbadian Dollar	English	279,200
Curaçao	Dutch Antilles/ Holland	Southern Caribbean	Netherlands Antillean Guilder	Dutch/English	192,000

Europe and the Mediterranean

While the Caribbean has benefited from the changing pattern of cruising in the aftermath of 9/11, Europe and the Mediterranean (Figure 4.3) are poised to develop exponentially as tensions concerning travel begin to ease (Wild and Dearing, 2004b). Barcelona and Palma in Spain and Venice in Italy lead the rankings list of most visited ports in southern Europe, reflecting a trend for itineraries to be located more toward the west of the Mediterranean or the Adriatic (Wild and Dearing, 2004c). Southampton in the UK has emerged as a leading port for the northern region, being appropriately located to service a diversity of itineraries and, according to Wild and Dearing (2004c), it is close to the types of facilities that allow effective service of passenger and cruise ship needs.

Northern Europe

This cruise region has a number of advantages. For US passengers it offers familiarity with the culture, the geography, and the attractions offered by the major cities (Mancini, 2000). For European passengers it provides an easy departure from home ports. The countries and ports are, for the most part, highly sophisticated (Cruise Europe, 2005) and able to cope with the complex demands that accompany the arrival of the largest of cruise ships. A number of cruise brands, such as the Cunard

Figure 4.3: Europe and the Mediterranean

Line and P&O Cruises in Southampton, have traditional roots in this region. Indeed, historically the UK is the home of cruising.

The season for cruising in northern Europe is relatively short, but the ports are popular so traffic can be concentrated for the short cruising season (Wild and Dearing, 2004c). Wild and Dearing (2004c) note that the majority of passengers for this type of vacation are likely to be first, North American passengers; secondly, UK passengers; and thirdly, German passengers. When marketing cruises in Northern Europe, cruise companies can focus on the British Isles, the Baltic, Iceland, the Arctic and the North Cape, the Norwegian fjords, and Western Europe (Wild and Dearing, 2004b: 17). The following describes key destinations from this area.

Southampton

Southampton is a city with a long maritime heritage. The city has experienced both growth and decline because of the historical development associated with shipping in general and the cruise industry in particular. It is currently experiencing growth. The port is well located for London, and it has excellent transport links and the infrastructure to service cruise ship needs. The port provides a launch pad for ships to travel with or across the Atlantic, to the Mediterranean, to the ports of northern Europe. Because of this, it is listed by Wild and Dearing (2004c) as the first-ranked port in northern Europe.

Helsinki

Helsinki is the capital of Finland. From a cruise perspective, the city is located in a strategically convenient part of the Baltic for itinerary planning. It is a bustling port with as many as 40 ferry departures daily at the height of season. The port is attractive and closely located to the city center (Cruise Europe, 2005). Finland is different from other Scandinavian countries for two reasons. First, it has a different language that is more similar to Russian and Estonian than to the languages of the neighboring Scandinavian countries. Secondly, it shares a border with Russia, which has resulted in Finland possessing a different history and culture. With two thirds of the country covered by forest and one tenth made up of inland lakes, the Finnish Tourist Board emphasizes nature and the environment (Boniface and Cooper, 2005). The port is a secondary base port as well as a port of call, or destination.

Copenhagen

Copenhagen is the capital of Denmark, the smallest Scandinavian country, and, after Helsinki, it is the third most visited port in northern Europe (Wild and Dearing, 2004c). Copenhagen has a reputation for having lively nightclubs and bars and is a major cultural destination (Boniface and Cooper, 2005). It is the home of the Carlsberg brewery, which is both a tourist attraction and working production center, and the world famous Tivoli Gardens, which is Europe's oldest amusement center. Much is made of the figure of the Little Mermaid, a statue in the harbor area representing a character from Hans Christian Andersen's stories. The city was the recipient of the World Travel Award as Europe's leading cruise ship destination in 2004 (Cruise Europe, 2005).

St. Petersburg

St. Petersburg in Russia has seen major growth in numbers of cruise passengers over the last five years and it was slated for 313 calls in 2004 compared to 263 in 2003. The city is said to be the most beautiful in Russia (Boniface and Cooper, 2005). The attractions it offers are both historical and cultural, including the Hermitage Museum, a former Tsar's palace; the Maryinsky Theatre, home to the world-renowned Kirov Ballet; the former summer residences of the Tsars, located on the outskirts of the city; and St. Isaac's Cathedral, the largest church in Russia (Cruise Europe, 2005).

Tallinn

Tallinn is a United Nations Educational, Scientific, and Cultural Organization (UNESCO) heritage site. It is the capital of Estonia and boasts what is said to be one of the few examples of an old city that has been kept intact. Tallinn has a history as a port that can be traced back to the tenth century, and evidence suggests there was a settlement on the site as long as 3,500 years ago. The city offers a number of attractions including parks, heritage buildings, palaces, and museums (Cruise Europe, 2005). Tallinn is the fifth most visited port in northern Europe (Wild and Dearing, 2004c).

Stockholm

Stockholm is the sixth leading northern European port (Wild and Dearing, 2004c). It is the capital of Sweden, a country that has the largest unspoiled wilderness in Europe (Bonifac and Cooper, 2005). The city of Stockholm is located on a number of interconnected islands at one end of Lake Mälaren. The city offers a broad range of attractions such as museums, royal palaces, and heritage attractions. The city itself is attractive to visitors, with its narrow pedestrian streets, good shopping and restaurants.

The northern European region may have a short season because of frequent inclement weather patterns from late autumn through to early spring, but the ports are popular and tourist-friendly. Whether a cruise is seeking the "land of the midnight sun" while cruising past the fjords of Norway or the northern lights of Aberdeen on the northeast coast of Scotland, passengers have many opportunities for memorable moments.

Table 4.5: **Northern Europe destination facts**

Destination	*Country*	*Region*	*Currency*	*Language*	*Population*
Southampton	United Kingdom	Northern Europe	Pound sterling	English	217,500
Helsinki	Finland	Northern Europe	Markka	Finish	1,163,000
Copenhagen	Denmark	Northern Europe	Danish Kroner	Danish	1,785,000
St. Petersburg	Russia	Northern Europe	Ruble	Russian	4,645,000
Tallinn	Estonia	Northern Europe	Krooni	Estonian	430,000
Stockholm	Sweden	Northern Europe	Swedish Kroner	Swedish	1,500,000

Southern Europe

In cruising terms, this region encompasses the eastern and western Mediterranean and provides access to a range of countries from a large number of ports. With long dry sunny summers, the Mediterranean climate is conducive to vacations (Boniface and Cooper, 2005). The region offers a great diversity of attractions, including historical sites, sophisticated cities, and beach playgrounds, all within relatively accessible cruising parameters (Mancini, 2000). The distances between ports and attractions allow cruise planners to schedule itineraries in this region to take advantage of the best timing and economical fuel consumption and, in addition, to take advantage of high-caliber supply networks.

The Mediterranean is popular with many cruise passengers. US passengers can take advantage of "grand-tour" approach to visiting Europe and facilitate border crossings, minimize language problems, and maintain a desired level of comfort. One drawback for US passengers can be the need to fly long distances to board the ship, although passengers are served by a multiplicity of arrival airports that provide easy access to base ports. Another factor that concerns some passengers is political unrest in countries and regions close to the Mediterranean. Passenger concerns about destinations are easily remedied by changing itineraries—a factor that has helped to generate the growth in popularity of cruising.

UK and European passengers have relatively easy access to the Mediterranean. P&O Cruises, Cunard, Saga Cruises, and other cruise brands operate a variety of cruises that depart from the UK. Other cruises that depart from the Mediterranean are usually between 1 and 2 hours flying time away from local airports. The season in the Mediterranean is being reappraised to stretch the shoulder periods (the months between high and low seasons).

Barcelona

Barcelona, in the western Mediterranean, is a Spanish city that has become the most visited port in the region (Wild and Dearing, 2004c). In addition to its status as a major base port, the city offers a broad range of attractions to make it a destination in its own right. The city is peppered with characterful architecture that was designed by Antonio Gaudí, and many tours visit his unfinished cathedral, the Sagrada Familia. The Ramblas provides a main walkway through the center of the city past the Barrio Gótico, the medieval core of old Barcelona (Boniface and Cooper, 2005). The port offers a contemporary setting for passengers to embark and disembark, with modern terminal facilities and network of services for passengers and cruise ships (Medcruise, 2005).

Palma, Majorca

Palma is also a Spanish city in the western Mediterranean. The island of Majorca is one of the Balearic Islands located off the southern coast of Spain. The other principal Balearic Islands are Ibiza and Minorca, which are also ports of call for cruise ships. Majorca is well known as a holiday destination, and, in recent years, the port has become a popular fly, cruise, and stay product (Medcruise, 2005). The island provides a variety of resorts and accommodation for this type of package. Palma, the capital of Majorca, is an attractive city that has a typical Spanish atmosphere, an impressive cathedral, a variety of shopping options, and close proximity to the beaches and other attractions.

Venice

Venice is actually in the Adriatic Sea, not the Mediterranean. This northern Italian city has had a long and turbulent history, and seems to be continuously struggling against the ravages of nature and time. Yet, in its unique setting, with its canals and car-free environment, Venice is special. The vast scale of a Grand class cruise ship drifting past St. Mark's Square beside antiquities such as the Doges Palace and the Basilica presents an incongruous sight. As a sea-based trading center, Venice has a maritime culture and has always made a living from the sea. Its excellent terminal facilities provide a point of arrival and departure and easy access to this attractive destination (Medcruise, 2005). Recently, proposals have been made to construct a tidal barrier to counter flooding problems caused by a combination of the city sinking into the lagoon (2 cm in 100 years) and rising tides (*BBC News*, 2003).

Naples

Naples is located in Italy, just to the south of Rome. The city is overshadowed by the ominous presence of Mount Vesuvius. This slumbering giant of a volcano provides a most impressive backdrop to Naples and is responsible for creating two of the area's attractions—the excavated Roman ruins of Pompeii and Herculaneum. The port gives easy access to the vast city, which can appear both lively and chaotic. This is the Mediterranean's fourth most visited port after Barcelona, Palma, and Venice (Wild and Dearing, 2004c).

Civitavecchia

This unfamiliar Italian port provides the gateway into Rome. The city of Rome is a "must see" destination for travelers to Europe. The city boasts a veritable cornucopia of classical ruins and architectural gems, including the Forum, the Colosseum, the Vatican, and St. Peter's Square, all within a modern metropolitan setting. Getting from Civitavecchia to Rome usually involves a taxi or coach journey, although the town also has a train station, which provides regular and easy connection. The port is a large sprawling area, and ships can be located quite far from the port gate. This distance can be traversed by a coach link or shuttle service, or taxi service, to the town center.

Savona

Savona in Liguria, northern Italy, is the seventh most visited port in the Mediterranean area (Wild and Dearing, 2004c). Costa Cruises, one of Carnival Corporation's cruise brands, has leased the modern terminal building in the city and makes good use of the facility to support its operations. Savona is in the heart of the Italian Riviera, a region of pretty seaside towns, spectacular coastlines, and a wide range of attractions.

Livorno

Livorno is a large, bustling port that services the surrounding region of Tuscany in Italy by providing a focal point for cargo, ferry, and cruise traffic. The cruise terminal is approximately a third of a mile from the city center. However, for many passengers that may be irrelevant because a key attraction is the city of Florence, which is approximately 55 miles (88 kilometers) from the port. The port also provides access to the beaches of the area, the famous wine region (Tuscany is well known for wines, including its famous Chianti), and many other attractive towns such as Pisa, Lucca, San Gimignano, Volterra, and Siena.

Dubrovnik

Dubrovnik is a major Croatian city and port. Despite suffering heavy shelling during the Serb-Croat war in 1991 and 1992, this famous old walled city has been completely restored to enable visitors to experience its atmospheric street scenery. Dubrovnik offers contrasting experiences to visitors. The city holds much interest, with its ancient walled ramparts and fortresses, narrow pedestrian lanes, and historical town buildings. The surrounding countryside and coastline provide a rich mix of geography, culture, and leisure activities.

Piraeus

Piraeus is a Greek port that may be seen by some as the Civitavecchia of Athens. Yet Piraeus has long been the gateway to Athens, and, as a result, it has a lively and bustling character. The harbor area is large and accommodates a diversity of shipping traffic, such as cruise ships, cargo vessels, and the ferries and hydrofoils that connect Athens and the mainland to the many outlying Greek islands. The 2004 Olympics led to considerable investment in the infrastructure of Athens and surrounding areas. Athens is another key destination for cruise passengers. The city has many treasures that attract visitors, including the Acropolis, the Parthenon, and the Agora, or marketplace (Boniface and Cooper, 2005). Athens can be reached from Piraeus by taxi, public bus, tour coach, and subway.

Santorini

Santorini is a Greek island in the Cycladic island group in the Aegean Sea, some 130 miles from Piraeus. The island offers spectacular scenery from its highest point, across the sweeping curvature of the crescent-shaped landmass out to sea. The island was originally a volcano, but when part of the volcano collapsed into the sea the unique terrain was formed. Some claim that Santorini was the setting for the lost city of Atlantis.

Rhodes

Rhodes, named after the phrase "The Island of Roses," is an attractive Greek Island where ancient history combines with the contemporary beach and sunshine holiday. Rhodes is also the name of the capital city, which today presents itself as a medieval old town with strong historical connections to the Knights of St. John. (The Knights of St. John was a religious and military order that was originally founded in seventh century BC to participate in war in the Holy Land.)

Mykonos

Mykonos is another Greek island in the Aegean Sea. This small island, with a population of just 15,000, transforms during the summer, when 800,000 tourists inhabit the hotels, guesthouses, and tourist accommodation. Mykonos's charm is the appearance of the main town, with its winding back streets, white painted buildings, and beautiful island scenery.

The Mediterranean possesses many special destinations and ports that are worth a visit. The attractiveness of this region, as well as the historical and cultural values, creates a strong lure for a broad range of cruise tourists. There are those who may have ancestral links to the area, others who seek learning and cultural enrichment, some who are attracted to the beauty of the scenery and countryside, and people who enjoy the climate. Invariably, there are many who seek a combination of these features.

Table 4.6: **Southern Europe destination facts**

Destination	*Country*	*Region*	*Currency*	*Language*	*Population*
Barcelona	Spain	Western Mediterranean	Euro	Spanish	1,500,000
Palma	Majorca	Western Mediterranean	Euro	Spanish	325,000
Venice	Italy	Adriatic	Euro	Italian	63,000 (central)
Naples	Italy	Western Mediterranean	Euro	Italian	1,000,000
Civitavecchia	Italy	Western Mediterranean	Euro	Italian	50,100
Savona	Italy	Western Mediterranean	Euro	Italian	62,000
Livorno	Italy	Western Mediterranean	Euro	Italian	170,000
Dubrovnik	Croatia	Adriatic	Kuna	Croatian	43,770
Piraeus	Greece	Eastern Mediterranean	Euro	Greek	182,671
Santorini	Greece	Eastern Mediterranean	Euro	Greek	10,000
Rhodes	Greece	Eastern Mediterranean	Euro	Greek	100,000
Mykonos	Greece	Eastern Mediterranean	Euro	Greek	11,000

North America

North America provides a number of embarkation points where US customers can join cruises and where overseas customers can join as fly-cruise passengers. In some cases these ports are home ports for US vessels and cruise companies. In addition, cruise itineraries can be constructed from US and Canadian ports to meet passengers' needs for cultural and geographical attractions. Irrespective of a cruise ship's flag or country of registration, because of the actions of US port health officials there

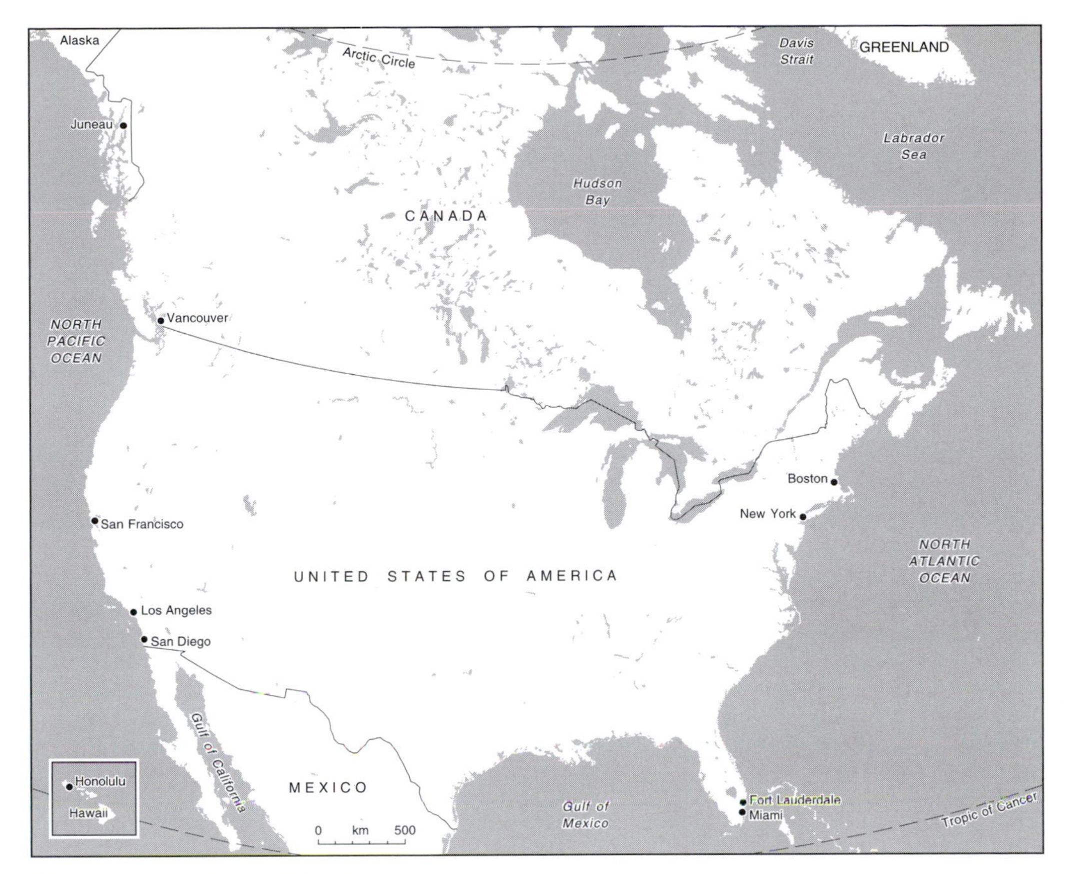

Figure 4.4: United States of America

are critical implications from the standpoint of port health for ships visiting US ports (which are examined in depth elsewhere) (Figure 4.4).

North America is the world's largest cruise market. As a result of concerns about security abroad, the number of cruise passengers joining from US ports has seen considerable growth. Of the top five cruise destinations in the world (Wild and Dearing, 2004a), Miami and Port Everglades are as number 1 and 2, respectively, and Port Canaveral is number 4. All three ports are in Florida. Cruise companies have benefited by consolidating their operational support in the United States and creating economies of scale from supply networks for merged brands. In addition, cruise companies have become horizontally and vertically integrated in their operations. Within the cruise industry, horizontal integration is attained by using different brands strategically within a variety of market segments. Vertical integration is achieved by creating synergies and generating revenue from ownership of parallel operations such as shore excursions, travel agents, terminal operations, and so on. The following list provides a brief outline of the features of major North American ports.

Miami

The port of Miami, on Dodge Island, is the busiest home port in the United States. As such, it provides a home base to Carnival Cruise Lines, Norwegian Cruise Line, Royal Caribbean International,

Oceania Cruises, and Windjammer Barefoot Cruises. The port has state-of-the-art facilities and has eight terminals with designated berths that can be used flexibly depending on the type of shipping. In addition to hosting passengers who are embarking on cruises, the port also provides facilities for cruise passengers arriving in Miami, with many options for excursions. It can also host those arriving in the city a day or so before departure (Port of Miami, 2005). While cruises can depart for many places, the main target is the Caribbean.

Port Everglades

Port Everglades is located close to Fort Lauderdale Airport, making for a relatively easy transfer for fly-cruise passengers who are primarily cruising to the Caribbean. Port Everglades, the number 2 port in the world, hosts many cruise brands, including Carnival, Celebrity, Costa, Crystal, Cunard, Holland America, Imperial Majesty, Mediterranean Shipping, Orient, Princess, Radisson Seven Seas, Regal, Royal Caribbean International, Royal Olympic, Seabourn, and Silverseas. The port provides a breadth of shore excusions (see Chapter 5) in the area (Port Everglades, 2005).

Port Canaveral

Port Canaveral has six cruise terminals with another two under construction. The port is home to Carnival Cruise Lines, Disney Cruise Line, Royal Caribbean International, Sterling Casino Lines, San Cruz Casino, Holland America, and Norwegian Cruise Lines. The port is in what is known as Florida's Space Coast, and visitors can take the opportunity to tour the Kennedy Space Center or indulge in a range of other activities (Port Canaveral, 2005).

Juneau

The port of Juneau provides access to the seasonal (May to September) attractions of Alaska. Juneau was a gold-rush town that became Alaska's capital. As a cruise destination, the city provides opportunities for exploring the area's mining heritage, participating in outdoor pursuits, visiting glaciers, whale watching, and even dog sledding. Glacier trips are available by helicopter. The city is more than simply a departure point for environmental pleasures, boasting an air of sophistication with its many art galleries and quality restaurants. Most of the major cruise brands that are marketed to US passengers sail to Juneau.

Ketchikan

Ketchikan is Alaska's southernmost city. Despite a high average rainfall, many outdoor pursuits are available, including kayaking, trekking, and visits to national parks, lakes, and forests. The city is a center for native culture, with an array of related museums and attractions.

Los Angeles

Los Angeles was the original home for the *Love Boat* television series that ran between 1977 and 1986. LA is famous for its many attractions, including Hollywood, Disneyland, and Universal Studios. The World Cruise Center in LA can manage a visit by the largest cruise ships (Cruise the West, 2005).

Long Beach

Long Beach is fast approaching the scale of operation at neighboring port Los Angeles (Wild and Dearing, 2004a). Carnival Corporation has a terminal at this port, and many cruise brands use Long Beach as a departure and home port. Itineraries from this port can include Baja California, the Mexican Riviera, and Alaska.

Tampa

Tampa handles a quarter of the numbers of passengers that Miami does (Wild and Dearing, 2004a). However, Tampa in Florida is expanding rapidly and attracts many of the leading cruise brands, including Carnival Cruise Line, Holland America Line, Royal Caribbean Cruise Line, and Celebrity Cruises. The port has a well developed, tourist-friendly, downtown waterfront area, and many excursions are available to augment the passenger experience.

A host of other ports lie within this large area, including Vancouver, New Orleans, Galveston, Skagway (Alaska), New York, New Jersey, Boston, San Francisco, Galveston, Philadelphia, and Seattle. Competition is fierce and growth, combined with recent trends, means many ports are experiencing "boom" conditions (Mott, 2004).

Table 4.7: **North America destination facts**

Destination	*Country*	*Region*	*Currency*	*Language*	*Population*
Miami, Florida	United States	North America	US Dollar	English	3,876,000
Port Everglades, Florida	United States	North America	US Dollar	English	40,000
Port Canaveral, Florida	United States	North America	US Dollar	English	15,000
Juneau	United States	Alaska	US Dollar	English	30,850
Ketchikan	United States	Alaska	US Dollar	English	8,000
Los Angeles, California	United States	North America	US Dollar	English	3,800,000
Long Beach, California	United States	North America	US Dollar	English	461,500
Tampa, Florida	United States	North America	US Dollar	English	303,500

Oceania and the South Pacific

Oceania, including Australasia (Australia, New Zealand, and Asia) and the islands of the Pacific, is a major expanse of sea and land. This cruise region offers great diversity from the culturally vibrant and exotic ports of Asia, such as Indonesia, Malaysia, the Philippines, Singapore, Thailand, India, Vietnam, China, Hong Kong, Japan, Sri Lanka and the Maldives, to the tropical islands of the Pacific, such as Tahiti, Fiji, Papua New Guinea, New Caledonia, Vanuatu, Samoa, Tonga, and the Cook Islands. Australia offers the attractions of her cities of Sydney, Melbourne, and Freemantle and the uniqueness of the coast, the coastal resorts, and the countryside (Cruise Down Under, 2004). New Zealand is still basking in the *Lord of the Rings* effect, which has followed the success of the movie trilogy. Publicity from feature films has frequently created heightened interest in travel destinations, and this has also been evident in the increased number of ships visiting the ports of Auckland and Wellington in New Zealand.

This region is located in the southern hemisphere, so the seasons are a reversal of the pattern recognized in the northern hemisphere. As a result, the summer cruising season for Australia and the South Pacific extends from November to April (Mancini, 2000). This vast geographical area is likely

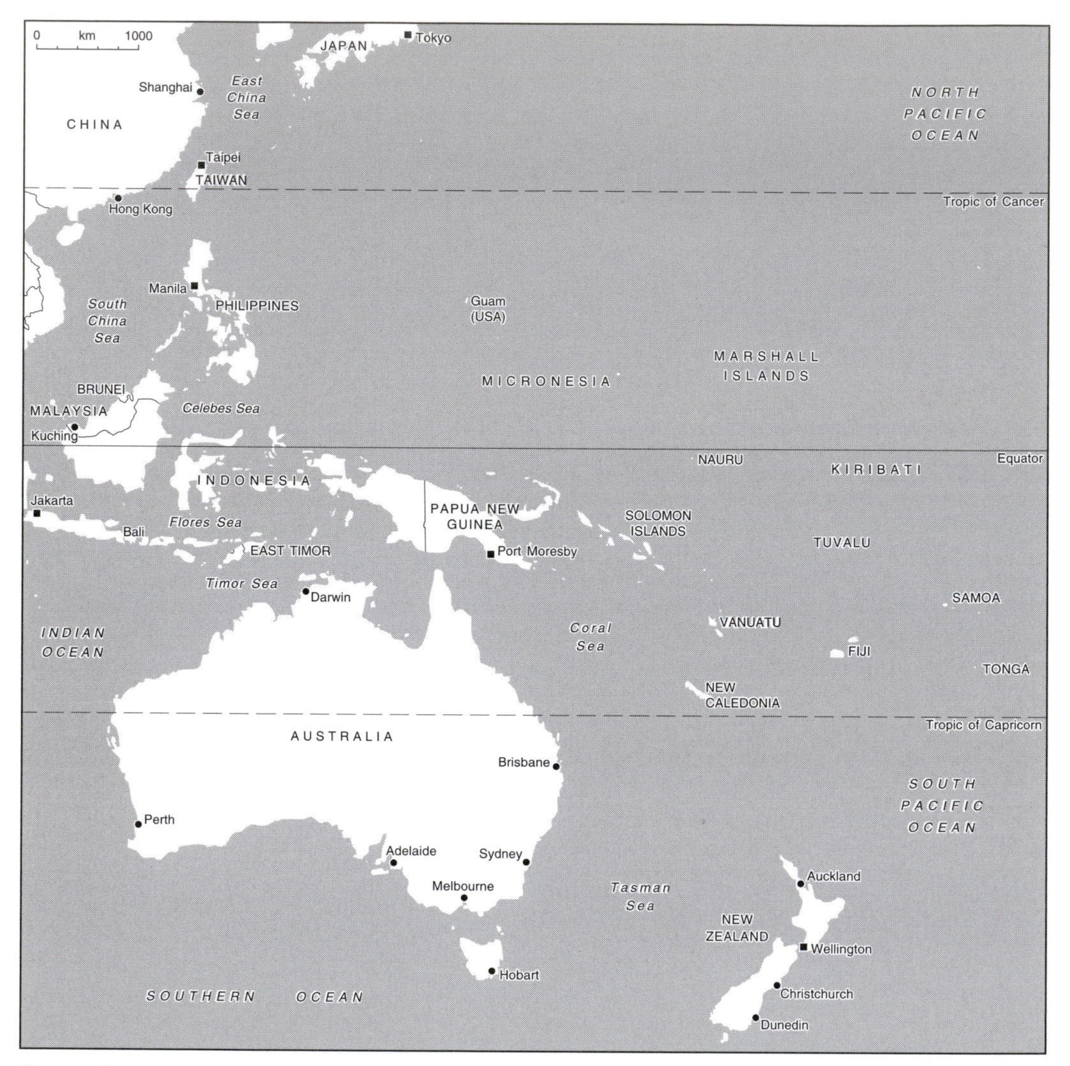

Figure 4.5: Oceania and the South Pacific

to experience continuous growth, with emerging economies such as China and India fuelling opportunities for new consumer markets and new, relatively accessible itineraries (Figure 4.5).

Sydney

Sydney is probably the best-known city in Australia, although it is not the capital (which is Canberra). The city has a highly picturesque setting, with its harbor, the Sydney Opera House, and the Sydney Harbor Bridge, which were blatantly and successfully exposed to the world media during the Millennium New Year celebrations. Sydney has two cruise terminals—the Overseas Passenger Terminal at Circular Quay and the Wharf 8 Darling Harbor Passenger Terminal—that are in close proximity to the city's attractions (Sydney Ports, 2005).

Auckland

Auckland is the largest city in New Zealand's northern island. It has an idyllic setting surrounded by islands and beautiful scenery. The city is both cosmopolitan and close to nature, with tours available to volcanic regions, rainforests, and beaches and the city's attractions within easy reach for passengers. The city has Polynesian and Maori culture, which is reflected in the people, the place names, the history, and the heritage of the area.

Fiji

Fiji is a group of 300 islands in the South Pacific with a population of approximately 893,000. The largest two, Viti Levu and Vanua Levu, contain 80% of the country's population. The islands represent many peoples' vision of what tropical islands should be like. There are white beaches, coral reefs, and clear seas with a myriad variety of fish and sea-life, alongside rainforests and native villages. Contemporary Fiji also has attractive shopping facilities in the capital, Suva, along with modern hotels, a wide variety of restaurants, and nightlife for all types of tourists.

New Caledonia

New Caledonia is a French island that lies halfway between Australia and Fiji. The island, with its capital Noumea, is the third largest in the Pacific after New Zealand and Papua New Guinea. The French cultural influence, coupled with the influence of the Melanesian region in this part of the Pacific (Melanesia is the name given to the island group that New Caledonia is part of), creates an interesting backdrop for this island. The geography is a mix of tropical features, with attractive sandy beaches, a large lagoon that surrounds the region, mountains, and rain forests.

Hong Kong

Hong Kong was formerly a British protectorate, but it was returned to China in 1997. It retains a mix of Eastern and Western influences and a dynamism that reflects a city on the cutting edge of a changing world. The city promotes itself as a shopper's paradise, but, in reality, there is more to this energetic, self-styled "cruise capital" of Asia. To the Western tourist a visit to Hong Kong by cruise ship is a special opportunity to savor its unique blend of sights and sounds (Hong Kong Tourism, 2005).

Singapore

The republic of Singapore is one main island surrounded by 63 smaller islets. It is an economically successful country, which is proud of its contemporary feel, its diverse culture, and its friendliness. Visitors can experience gardens, skyscrapers, the famous Raffles Hotel (home of the Singapore Sling), a strong sense of fashion, and a technologically aware community.

Table 4.8: **Oceania and South Pacific destination facts**

Destination	*Country*	*Region*	*Currency*	*Language*	*Population*
Sydney	Australia	Australasia	Australian Dollar	English	3,879,400
Aukland	New Zealand	Australasia	New Zealand Dollar	English	367,700
Fiji	Fiji	South Pacific	Fijian Dollar	English	893,000
New Caledonia	New Caledonia	South Pacific	*Comptoirs Francais du Pacifique franc* (XPF)	French	216,400
Hong Kong	China	SE Asia	Hong Kong Dollar	Chinese (Cantonese), and English	6,898,600
Singapore	Singapore	SE Asia	Singapore Dollar	Mandarin, English, Malay, Hokkien, Cantonese	4,425,700

Other Cruise Destinations

The capsule descriptions of the aforementioned cruise regions are, of course, flawed because they only scrape the surface of the options for creating a cruise itinerary. Further research will undoubtedly reveal a plethora of destinations that are unmentioned in this chapter but, nonetheless, hold vital importance as part of a cruise itinerary. At the end of this chapter, the reader will find links that can help with such research.

Some additional destinations are worth highlighting and have until now been omitted because of their geography. For instance, nothing is said about most of Africa and the islands off Africa, such as the Canaries, Mauritius, and the Seychelles. South America is also omitted, despite its obvious attractiveness and a wealth of interest in the continent. A few ports from these areas are mentioned here to draw attention to their potential.

Atlantic Islands

A number of primarily volcanic islands form part of cruise itineraries in the Northern Hemisphere. These are the Canary Islands, Madeira, and the Azores. The Canary Islands of Tenerife, Lanzarote, Gran Canaria, and Fuertenventura are governed by Spain, although they fall outside the jurisdiction of the European Union (EU), which means that cruise ships with an EU registration can have the opportunity to sell duty-free alcohol. An EU-registered ship that has an itinerary made up of destinations or ports that are all EU member states would not be able to make such sales. Madeira and the Azores are Portuguese islands. The Canaries are relatively close to the coastline of North Africa and benefit from a temperate climate all year round. Madeira has a similar climate and is a popular cruise destination for passengers who enjoy the verdant scenery and charm of Funchal, the island's capital. The Azores offer a different type of destination. The islands have a quietness about them that reflects their remote setting. These islands were a convenient stopping point for trans-Atlantic crossings, but are now less frequented because the need for such a logistically convenient stopover is much reduced (Boniface and Cooper, 2005).

Rio de Janeiro

This major Brazilian city conjures up images of Sugarloaf Mountain, with its world famous statue of Christ facing over Rio's population. Copacabana Beach and Ipanema Beach are also well known as playgrounds for locals and tourists alike. The city is representative of Brazilian exuberance, as seen in the everpresent music, the dancing, and festivals.

Buenos Aires

Buenos Aires is the capital of Argentina. It is an optimistic and proud city with a much publicized past. Its architectural heritage is European, with influences from Britain, France, Italy, and Spain. Its museums, theaters, and art galleries reveal the cultural proclivities of the locals. Much is made of the links between the nation and the tango, a dance that embodies passion and drama (Boniface and Cooper, 2005).

The Galapagos Islands

The Galapagos Islands are Ecuadorian islands in the Pacific. Despite being almost barren, these small islands are popular cruise destinations because they present an ecosystem that is unique. The water is cold, yet the islands lie on the equator so the mix of land and sea creatures is diverse. There are no natural predators, so the indigenous animals, including giant tortoises, marine iguanas, penguins, and sea lions, have no built-in fear of humans. The area is extremely sensitive, and cruise ships and passengers are managed with great care (Boniface and Cooper, 2005) to minimize environmental impacts.

Cape Town

Cape Town is the capital of South Africa. The city is located in a place where cold and warm collide, because of sea currents from both the Atlantic and the Southern Oceans. Cape Town is a natural harbor that makes an excellent destination with a wealth of options for the cruise tourist, including the friendly locals, the dramatic scenery of Table Mountain, the famous regional vineyards, and the beaches.

Seychelles

The Seychelles in the Indian Ocean are idyllic islands that are truly beautiful. The clean sandy beaches, clear seas, palm trees, and granite outcrops peppering the shores offer a relaxing port of call for cruise itineraries. The flora and fauna on these islands are unique because of their distance from the nearest landmass. Cruise tourists are most likely to visit Mahe, the main island.

Panama Canal

The Panama Canal is not really a destination in its own right, but it is vital as a link between the Atlantic and the Pacific Oceans and a fascinating experience for cruise passengers. The first ships used the canal in 1914. From then it was operated by the United States until 1999, when it was returned to the Panamanian government. It has three sets of locks to facilitate the different sea and water levels, and, on average, it takes about 8 to 10 hours for vessels to get from one side to the other. Ship dimensions must not exceed 32.3 meters in beam, draft 12 meters, and 294.1 meters long (depending on the type of ship).

Suez Canal

The Suez Canal started operations in 1869 to provide a link between the Mediterranean and the Indian Ocean. Using this shortcut, ships could avoid the potentially dangerous and lengthy voyage around the Cape of Good Hope. Since that time, with occasional closures at time of war, the canal has become on of the world's most important trading routes (Boniface and Cooper, 2005). The canal can allow vessels up to 150,000 GRT with a 15-meter draft to traverse the canal, although plans are under way to increase this limit to a 20-meter draft by 2010.

Table 4.9: **Other destination facts (population figures are approximate)**

Destination	*Country*	*Region*	*Currency*	*Language*	*Population*
Canary Islands	Spain	Atlantic	Euro	Spanish	1,672,600
Madeira	Portugal	Atlantic	Euro	Portuguese	245,000
Azores	Portugal	Atlantic	Euro	Portuguese	237,000
Rio de Janeiro	Brazil	South America	Real	Portuguese	5,093,000
Buenos Aires	Argentina	South America	Argentine Peso	Spanish	11,928,000
Galapagos	Ecuador	Pacific	US Dollar	Spanish	18,000
Cape Town	South Africa	Africa	Rand	IsiZulu, IsiXhosa, Afrikaans, Sepedi, English	3,092,000
Seychelles	Seychelles	Indian Ocean	Seychelles Rupee	Creole and English	81,100
Panama Canal	Panama	South America	The Balboa or US Dollar	Spanish and English	3,039,000
Suez Canal	Egypt	Africa	Egyptian Pound	Arabic	469,500

Summary and Conclusion

This chapter provides a description of the major cruising sectors and a brief taste of a number of destinations. Considerably more can be said about all the destinations that are included, as well as those that are not. However, this is not possible in such a broad-based textbook, and it is recommended that readers undertake further research to examine key issues related to destinations and cruise sectors. Many good resources exist, including Web-based tourism sites, geography textbooks, and tourism guides.

The cruise industry generates considerable business for destinations, but for some there is a cost. That cost may be in terms of the increase in people visiting particular destinations, the demands placed on the local population to "package" and thus taint the cultural experience, the possibility of pollution, or ecological impact. These issues can also be examined further with respect to the balance of the positive and negative impacts on destinations.

The following two case studies are included to stimulate discussion and widen understanding about destinations. The first is the case of "Destination Southwest," a regional initiative that was introduced to increase cruise business to the southwest of England. The second examines the action taken by some cruise companies to purchase and operate private islands. Questions are included at the end of each case study.

Case Study: Selling the Southwest of England

Resorts seeking to capitalize on the burgeoning cruise phenomenon could learn by examining the case of "Destination Southwest." This initiative, representing an alliance among eight ports throughout the southwest of England (Ilfracombe, Torbay, Dartmouth, Plymouth, Fowey, Falmouth, Penzance and the Isles of Scilly) was supported by a combination of European Regional Development Social Funds and match-funding from county and local councils, tourists bodies, and attractions such as the Eden Project, the National Maritime Museum in Cornwall, the National Trust, Britannia Royal Naval College, and the ports (a total of 21 public and private partners). The funds helped support a new partnership that aimed to develop and extend the number of cruise ship visits to ports within the region.

Bob Harrison, an experienced professional with 30 years of industry experience, coordinated Destination Southwest at sea and ashore. He was appointed director of cruise operations. Harrison used his knowledge and contacts to gain access to senior managers involved with itinerary planning, and his insights into the experience and requirements of cruise passengers have been invaluable. He also recognized the difficulties that cruise executives face when planning itineraries and can orient his strategy accordingly.

The £230,000 three-year project was launched in February 2002. Destination Southwest started by establishing an informative website, www.destinationsouthwest.co.uk. The website allowed visitors to click on the port name and get access to a lot of information, including a cruise calendar to identify which ships are calling and when, marine charts, town maps, suggested shore excursions, video clips of some attractions, 360-degree shots of the port area where passengers land, and distances and times between ports and attractions. This information, which helps the itinerary planner make decisions, was also replicated on a DVD to provide an easy-to-use reference in support of direct selling and to give away at exhibitions. Websites and DVDs were produced in German and English to appropriately target the US and German cruise markets.

There were a few problems. Events such as the terrorist attack on New York, the outbreak of foot and mouth disease in the UK in 2003, and the emergence of SARS all affected purchasing decisions by potential cruise passengers. In many cases, people were ill informed about the implications of these critical incidents. Passengers to a port in Cornwall were overheard asking whether purchasing a woolly pullover might create a risk of catching foot-and-mouth disease.

Cruise companies such as Holland America, Princess, Cunard, and Seabourne all liked the uniform packaged approach. The results of the project show an increase from 10 vessels calling into ports in the area in 2001, to 106 cruise ship calls in 2004. By all accounts, this is a dramatic increase. Some ports in this region offer berths where the ship can go alongside. Indeed, the busiest port, Falmouth, has such facilities. Harrison believes that passengers prefer as few tender operations as possible (where launches transport passengers from ship to shore). This despite the fact that many cruise brands include a tender operation as part of the total cruise experience. In his opinion, more than two tender operations in a cruise is too many, because passengers begin to object to the time delay, the potential discomfort if sea conditions are not calm, and queues that can form at either end of the operation because of security and logistical factors. The business generated by Destination Southwest includes 21 turnarounds. A turnaround port is one where the cruise starts and finishes. Dartmouth is reported to operate one turnaround while Falmouth has 20. The Falmouth turnarounds involve a ship called the *Van Gogh*, which has itineraries to the Mediterranean, the Canary Islands, and the Caribbean.

Harrison describes recent research on passenger spending, which focused on Cork in southern Ireland. He quotes this Irish research because he believes that cruises based in Cork and those in the Destination Southwest program have much in common. This research identified that each passenger spends £197 and that crew members spend marginally less. There is, however, disagreement about spending levels, and some sources suggest that at times the crew can actually spend more than some passengers. This is explained by patterns in port manning. Minimum levels of crew must remain on board when the ship is in port to ensure safety. As a result, crew members cannot go ashore in every port but when they do go ashore they are more likely to spend greater amounts of money. Harrison estimates that cruise passenger spending in the southwest of England is £16.7 million, while crew spending is £1.7 million. That is a total of £18.4 million from a project with a £285,000 budget, which could equate to 438 jobs for the local community. When a cruise ship comes to Falmouth, the local department store, Marks and Spencer, takes on extra staff. The project has attracted high-profile vessels to the ports in the partnership and it is reasonable to highlight the benefit to the port's image. When the '*World*' was visiting Falmouth as part of her itinerary, there were 2,000 people standing on the headland to see the ship as she sailed out of the harbor.

How was this level of success achieved? The website was seen to be important. The quality of information and the ease of use are fundamental. Destination Southwest possessed a tacit understanding of the cruise industry and its requirements. The project used business-to-business (B2B) marketing. Personal contacts that, developed relationships and encouraged visits to the area by decision makers were also important. Attendance at trade and travel conventions was of importance because it gave Destination Southwest a presence. Douglas Ward, the author of the *Berlitz Guide to Cruising*, was appointed as honorary president, and this link was also thought to be useful because of opportunities to enhance networking and raise the project's profile. The initiative has helped to develop the level of support for cruise tours, quality of welcome for cruise passengers, and overall focus for customer service. In Torbay, for example, shop-mobility trolleys were made available, and the mayor attended personally to welcome the passengers. In other ports, such as Plymouth and Falmouth, portable tourist information display units were available for passengers to consult.

Local problems that negatively affect business include conflicting schedules from ferry operators competing for berths and limited under-keel clearance in port areas that inhibits ship mobility and access during certain tides and times. The project has helped lengthen the tourist season and has brought many people from the US, Europe, and the UK to the region. The potential for cruising is continually expanding. The Passenger Shipping Association states that there was a 37 percent increase in visits to the UK between 2003 and 2004 and identifies it as the fastest growing market in the world. The potential for the southwest of England is high. Even the weather, often regarded by some as a turn-off, is seen by many US passengers as an attractive experience. The local government office is impressed with the project and aims to extend it for a further two years. Levels of investment are low, however, and there is a constant struggle to persuade members to contribute. There are major benefits to acting in concert. Ports that do not act in competition can derive benefits through cooperation. Harrison says that cruising is for everyone, and the challenge is to inform those who do not recognize how it has changed.

The preceding case study presented an account of how one region in a country created a plan to develop and sustain cruise tourism growth.

Case Study Questions

1. Consider the key actions and identify the critical elements that led to the outcome of this initiative.
2. What are the risks for this project and how can they be addressed with respect to the following?
   a. Internal competition between neighboring ports
   b. Ensuring the ports remain attractive as cruise destinations
   c. Securing finance to ensure that the project develops

Case Study: Private Beaches as Ports of Call

Private beaches, such as those owned or leased by cruise brands in the Caribbean or the Bahamas, are seen as a useful alternative to neighboring popular ports of call. Often the beaches are on Cays, the local name given to small islands, which is a derivative of the word Key (as in Key West). Cays are small, low-lying islands consisting mainly of coral and sand. Cruise lines such as Disney, Princess Cruises, Norwegian Cruise Line, Holland America Line, Costa Cruise Lines, Royal Caribbean International (which has two islands), and Radisson Seven Seas Cruises are all involved in this type of investment.

But what are the advantages and are there any disadvantages in having a private beach port? Most of the islands are constrained by their location and facilities, thus requiring that the ship anchor off the coast, with passengers then ferried to the island jetty by tender. This transfer can add an exciting dimension to a cruise, although those with small children or with a disability may be inconvenienced. The notion of a private cay or beach can be attractive to passengers because of the implied romance or because the idea may signify to some a prestigious and unique benefit.

The visit to the cay is often scheduled to include a morning arrival and late afternoon departure. This optimizes usage of the cay and allows the company to build in additional services such as barbeques, water sports, and organized games and activities. This in turn creates opportunities to generate revenue for the activities and for facilities. Kayaking, sailing, snorkeling, scuba diving, and a range of children's activities can be scheduled. In addition, some companies have scheduled special activities such as massages in private cabanas (Disney and Holland America) and "surf and turf" Olympics (Costa Cruises). Royal Caribbean has built a replica of a Spanish galleon and sunk a small airplane in the waters off its Bahamian island, 140-acre Coco Cay, for snorkeling tours and scuba divers.

Services may be provided by the cruise company or subcontracted to local employees or contract providers. The services are under the quality assurance and control of cruise management with, in some cases, shipboard staff being used ashore to create a seamless service. In addition to this approach to developing their "products," cruise companies are also introducing "beach clubs" in popular destinations, which are managed and operated directly or as part of a contract by the cruise companies.

Some observers are critical of this approach (Robertson, 2004), noting that issues relating to the environment and the amount of waste generated by tourists, both on the ship and when visiting these fragile islands are important and in need of further examination. In addition, points are raised about the ethical position of cruise companies in relation to the playgrounds of the Caribbean and the Bahamas. Cruise companies generate large amounts of revenue from their islands by selling products and services but, it is claimed, in doing so, the direct contribution to locals trading in the Caribbean is being eroded (Robertson, 2004).

Case Study Questions

1. What are the significant advantages and disadvantages for the various stakeholders: the cruise company, the locals, the passengers, and the relevant authorities?
2. In some ways this approach to developing a resort is criticised as being an example of "enclave tourism," where tourists are sheltered from a local environment by barriers intended to protect the tourists and manage their experience (Boniface and Cooper, 2005: 453). Why is this so and what are the implications?

Glossary

Archipelago: A group of many islands.
Duty-free: The status claimed when goods are sold without the need to collect any state tax.
DVD: Digital video disc.
Gross Domestic Product: The total value of goods and services produced by a nation.
Port fees: Levies charged per passenger for a cruise ship entering a destination port.
SARS: Severe Acute Respiratory Syndrome.

Chapter Review Questions

1. What are the primary cruising sectors?
2. What are the important factors that define a port or destination?
3. What part does weather play for cruise destinations?
4. What are the secondary and emerging cruise sectors?

Additional Reading and Sources of Further Information

http://www.cruising.org/planyourcruise/resources.cfm—links to tourist organizations
http://www.cybercruises.com/cruiseports.htm—links to port authorities and consortia
http://www.port-of-call.com/Portsofcallv2/default.asp—French ports of call

References

Antigua Barbuda Tourist Information (2005), Antigua and Barbuda. Retrieved 13 April 2005, from http://www.antigua-barbuda.org/

Aruba Cruise Tourism (2005), Aruba Cruise Passengers. Retrieved 2 Aug 2005, from http://www.arubabycruise.com/stats/index.html

Bahamas Tourism Office (2005), Experience the Bahamas. Retrieved 13 April 2005, from http://www.bahamas.com/bahamas/

Bansal, H., and Eiselt, H. A. (2004), Exploratory research of tourist motivations and planning. *Tourism Management*, 25(3), 387–396.

Barbados Tourism Authority (2005), Barbados. Retrieved 13 April 2005, from http://www.barbados.org/

BBC News (2003), Venice launches antiflood project. Retrieved 2 Aug 2005, from http://news.bbc.co.uk/2/hi/europe/3026275.stm

BBC News (3 March 2004), Crisis in Haiti, from http://news.bbc.co.uk/1/hi/world/americas/3378671.stm

Boniface, B., and Cooper, C. (2005), *Worldwide destinations* (4th ed.). Oxford: Butterworth Heinemann.

Burton, R. (1995), *Travel Geography* (2nd ed.). London: Pitman.

Caribbean Tourism Organization (2005), Caribbean—everything you want it to be. Retrieved 11 April, 2005, from http://www.doitcaribbean.com/cruising/

Cayman Islands Department of Tourism (2005), Cayman Islands—close to home, far from expected. Retrieved 13 April 2005, from http://www.caymanislands.ky/

CIA (2005), Aruba: World Fact Book. Retrieved 2 Aug 2005, from http://www.cia.gov/cia/publications/factbook/geos/aa.html#People

CLIA (2005), Bahamas and the Caribbean. Retrieved 11 April 2005, from http://www.cruising.org/planyourcruise/wwdest/destination.cfm?ID=7

Cruise Down Under (2004), News. Retrieved 27 April 2005, from http://www.cruisedownunder.com

Cruise Europe (2005), Webpage. Retrieved 20 April 2005, from http://www.cruiseeurope.com/

Cruise the West (2005), Cruise partnership. Retrieved 27 April 2005, from http://www.cruisethewest.com

Curaçao Tourist Board (2005), Curaçao. Retrieved 13 April 2005, from http://www.curacao-tourism.com/

Dervaes, C. (2003), *Selling the sea* (2nd ed.). New York: Thomson.

Florida Keys and Key West Tourism Association (2005), Florida Keys and Key West. Retrieved 13 April 2005, from http://www.fla-keys.com/

Gibson, P. (2004), Life and learning in further education: constructing the circumstantial curriculum, *Journal of Further and Higher Education*, (28), 333–346.

Hong Kong Tourism (2005), All about Hong Kong. Retrieved 27 April 2005, from http://www.discoverhongkong.com/eng/mustknow/index.jhtml

Mancini, M. (2000), *Cruising: A guide to the cruise line industry*. Albany NY: Delmar.

Medcruise (2005), Cruising in the Mediterranean. Retrieved 23 April 2005, from http://www.medcruise.com

Mott, D. (2004), Home comforts. *Lloyd's Cruise International,* 17–19.

Port Canaveral (2005), Cruising from Port Canaveral. Retrieved 27 April 2005, from http://www.portcanaveral.org/

Port Everglades (2005), For travel professionals. Retrieved 27 April 2005, from http://www.sunny.org/travelagents/index.cfm

Port of Miami (2005), Cruise. Retrieved 27 April 2005, from http://www.miamidade.gov/portofmiami/cruise.asp

Puerto Rico Tourist Office (2005), Go to Puerto Rico. Retrieved 13 April 2004, from http://www.gotopuertorico.com

Rees, R. (1992), *Mitchell Beazley's family encyclopedia of nature*. London: Mitchell Beazley.

Robertson, G. (2004), Cruise ship tourism. Retrieved 29 April 2005, from http://www.lighthouse-foundation.org/

Sydney ports (2005), Sydney ports: first port, future port. Retrieved 27April 2005, from http://www.sydneyports.com.au/home.asp

US Virgin Islands Tourism Authority (2005), St. Thomas. Retrieved 13 April 2005, from http://www.usvitourism.vi/en/stthomas/st_Home.html

Visit Jamaica (2005), Explore Jamaica. Retrieved 13 April 2005, from http://www.visitjamaica.com/home/Default.aspx

Wild, P., & Dearing, J. (2004a), Caribbean Stronghold. *Lloyd's Cruise International* (69), 27–38.

Wild, P., & Dearing, J. (2004b), Growth culture. *Lloyd's Cruise International*, 17–24.

Wild, P., & Dearing, J. (2004c), Rising stars. *Lloyd's Cruise International.*

Wild, P., & Dearing, J. (2005, December–January 2005), High achievers. *Lloyd's Cruise International*, 19–28.

World Travel Awards (2004), Awards. Retrieved 20 April 2005, from http://www.worldtravelawards.com

5

Planning the Itinerary

Learning Objectives

By the end of the chapter the reader should be able to:

- Define the cruise destination
- Identify critical factors relating to the design of a cruise itinerary
- Examine different methods for analyzing and evaluating destinations
- Understand why ports and destinations are successful for cruises
- Consider operations and planning for shore excursions or tours

In the previous chapter, cruise geography and travel geography were examined (Burton, 1995) alongwith factors such as seasonality and optimum conditions for cruising. Sales and marketing and the supply and servicing of ships were also considered as factors that are important for designing itineraries. This section will examine options for analyzing destinations for itineraries.

What Is a Cruise Destination?

This is a relatively complex question to answer. Davidson and Maitland (1997) present a model that describes the interplay between a "generating region," the place that the tourist will come from, and "destination regions," the places tourists will go to, linked by a "transit region," the place where the tourist spends time before arriving at the destination. In this version of a tourist system, potential tourists within a generating region are subject to a variety of "push" factors, such as disposable income, leisure time, motivation, ambition, and the presence of demographic change. Information is channeled back to the generating region from the destination region, developing "local" perceptions and stimulating further visits.

For the cruise industry, noting the significant changes in recent years in terms of the construction of larger vessels with enhanced facilities, the key destination can be interpreted as being the ship itself. Indeed, the ship has a significant place in the cruise tourism system, as can be seen in the following modified model (Figure 5.1).

This reworked version of the tourism system suggests that the ship plays a pivotal role in the relationship between the generating region and the ultimate tour destination regions. In a sense, over the duration of a cruise, it becomes a center for interpretation, a safe and familiar zone from which it is possible to choose whether to select, sample, and engage with situated experiences. Some passengers prefer to stay on board during a port day, rejecting the attractions ashore in favor of the shipboard experiences. Information travels from the destination region to the ship and then from the ship to the

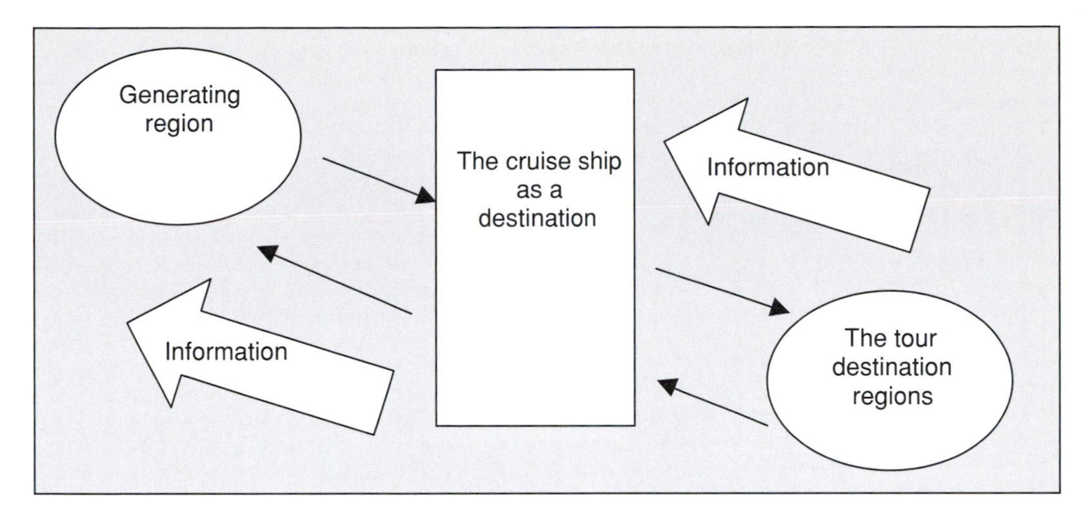

Figure 5.1: The cruise tourism system (after Davidson and Maitland, 1997)

generating region. Of course, this can be further developed when considering that some cruise companies have experimented with selling cruises to nowhere.

What Makes a Good Port of Call?

Ports of call, or destinations, invariably offer a mix of elements which, taken together, have potency. Cruise ships are businesses that rely on customer satisfaction. The main feedback from cruise tourists is obtained using a survey document that is distributed, completed, and returned at the end of the cruise.

Passenger feedback suggests that ports should be interesting, culturally stimulating, safe and non-threatening, friendly, accessible, and user friendly. It is difficult to find an ideal port that checks all these boxes, so in reality compromises are made and the difference between expectation and perception of actual experiences tends to stimulate positive reflection.

Ports of call derive considerable income from cruise ships and popular destinations sell themselves aggressively in order to attract cruise tourism. An analysis of port advertisements in *Cruise International* (Lloyd's, 2003a) reveals the following range of attractors (see Table 5.1).

A variety of marketing communications (trade magazines, direct selling) and forums (trade shows and conferences) are used by marketers to sell the benefits of destination ports. Increasingly, the Internet has become a powerful communications tool. Examples below and from the previous chapter show how individual ports and consortia approach the task of using the Internet as an aid to sales.

- www.cruisejamaica.com
- www.destinationuk.co.uk
- www.arubabycruise.com
- www.dubaitourism.co.ae
- www.marmariscruiseport.com

An analysis and comparison of the above sites will provide interesting lessons about how the Internet is used as a marketing tool for the cruise industry.

Table 5.1: **Analysis of attractors**

Unique experiences	*Heart of the city location*
Average 35 ft. (10.75 metres) water at low tide	Shopping
Deep draft sheltered berths	Capacity for megacruise ships
Gateway port with easy access to destinations	ADA-accessible passenger loading bridge and mobile gangway
Port an attraction in itself	Comfortable, efficient and secure
Duty free	Dual-ship terminal
Suitable as home port, port of call or repositioning cruise port	Warehousing space (storage, stores and baggage handling)
Professional service	Panoramic views
Island port with diversity of attractions	International airport nearby
Cruise terminal with state of the art facilities	Perfect weather year round/warm weather destination
Sightseeing tours/shore excursions	Cultural and historical treasures
Exciting nightlife	Water sports and land sports

Analysis and Evaluation

There are many analytical tools that can be used to measure the potential value of a port. Invariably, the decision is complex and takes into account many practical factors. For an established cruise brand, there is much to be gained from building on experience and planning itineraries based on what is known because of the reliability factor. For the cruise ship, experience with visiting a port creates a knowledge bank about the destination, which helps to ensure that planning is effective and that quality is less of an unknown commodity. For the port of call, experience enables agents, port officials, contractors, tourist organizations, and the local population to form and develop a relationship with the visiting ship and its community.

According to Lloyd's (2003b), a company such as Silverseas involves a broad group of stakeholders to compose the itinerary. This group includes captains, sales teams and passengers. Passenger questionnaire responses inform the process, as do world events. The company adopts an eclectic approach to itinerary planning to reflect the developing and ever-changing needs and wants of clients. Accordingly plans are incorporated to ensure that those with a phobia for flying are accommodated, new ports are included to ensure the itineraries are not seen as staid or lacking adventure, land based activities including cruise and stay programs are constructed, and details relating to passenger expectation are incorporated throughout. For Silverseas doing so involves focusing on the luxury end of the market with unique or prestige events included in itineraries to create special moments in keeping with customer expectations, such as a dinner in St. Petersburg museum or a private opera in the Sydney Opera House. Silverseas is reported to have reduced the duration of the average cruise from between 14 and 16 days to between 9 and 12 days in response to passenger wishes.

Problems related to itineraries can involve practical matters such as tendering—the ferrying of passengers by ship to shore using ship's tenders. Tendering is said to be less popular with many passengers because of a combination of factors. For some it increases travel time to land and reduces time ashore, for others it is an unwelcome form of transport that can raise concerns. On a more positive note, the experience can add to the total experience that the cruise provides by providing a frisson of excitement. Passengers who transfer from the grand scale of the cruise ship to the more human-scaled tender have an opportunity to enjoy a different mode of transport with a unique view of both the port (outbound) and the ship (inbound). Some ports of call have reputations about how they deal with visiting cruise ships. According to some captains there are ports where the authorities increase port costs without prior notice, where an expensive landing tax is levied and where a charge is made for compulsory but unnecessary tugs.

With respect to a ship's itinerary, the purpose of destination analysis is most likely to be strategic; that is, related to long-term corporate objectives and connected to creating sustainable competitive advantage. This focus invariably involves a strategic evaluation of the appropriateness of the destination for the target market and research data in deciding what makes a useful itinerary or component of an itinerary. Analysis can also inform tactical decision making, feeding into the overall strategy with a more medium-term concern, such as changing an itinerary because of emerging problems or heightened risk.

Analysis can came from different perspectives. For example, a destination can be evaluated by a tourist agency, a tour operator, or a cruise company. In this sense the analysis can be defined as being internal or external. Internal analysis reflects on issues such as strengths and weaknesses including core competence (those unique characteristics that say what a company does best), the tangible and intangible resources (the physical entities such as buildings or stock, skills and brand names), and financial aspects. External analysis considers the view outside of the study area but still important to it and considers opportunities and threats (Evans et al., 2003).

A general analysis of destinations can involve business considerations, such as maximizing financial returns, or a reflection of sociological perspectives to create a depth of understanding about the destination (Framke, 2002; Melian-Gonzalez and Garcia-Falcon, 2003). The former approach is more focused on opportunities and threats for a business venture in a pragmatic manner, whereas the latter may be intended to unearth social and cultural meanings that can be vital but also, in their own way, almost esoteric (providing information that is only useful to a specific group with specific knowledge).

The following section provides a range of analytical tools that can be applied in the task of itinerary planning. These tools can be used to look at the macroenvironment (the broad influences that can be said to affect the whole industry) or the microenvironment (the influences that immediately surround a business) (Evans et al., 2003). They can consider the destination as an attraction or consider tourists in terms of their needs and wants. The first example suggests a range of generic analytical approaches—those that can be applied to a variety of settings. Thereafter, variety of analytical approaches are suggested that have more direct relevance to tourism and destinations.

SWOT

The SWOT analysis (which stands for Strengths, Weaknesses, Opportunities, and Threats and is sometimes referred to as TOWS or TWOS) is the archetypal approach to a strategic evaluation. It embodies an internal analysis of strengths and weaknesses and an external analysis of opportunities and threats (Evans et al., 2003). It is possible to apply this analysis from the port or destination's perspective when considering inherent factors or it can form an important analytical tool from a cruise operator's perspective when examining a port or destination in the context of an itinerary. The SWOT analysis is frequently applied to historical factors that had a major impact in the past, current and nascent factors that are likely to affect on future performance, and factors that render the organization distinctive from the competition. The best SWOT analysis is one that is supported by logic, argument, and evidence.

PESTLE Analysis

This is a very common analysis that is used by organizations to study the external macroenvironment. The acronym stands for Political, Economic, Social, Technological, Legal, and Environmental.

Variants of this analytical tool include PEST, SPECLE, STEP, PEST, STEEP, SPECLE, SCEPTICAL: the acronyms can be spelled out by extracting from terms used in the next sentence. The final example, Social, Cultural, Economic, Physical, Technical, International, Communication and Infrastructure, Administrative and Institutional, Legal and Political, was constructed by Peattie and Moutinho (2000) with specific application to travel and tourism. This type of analysis has four stages: scanning to identify signs of risk or environmental changes, monitoring to recognize patterns and trends, forecasting to calculate future environmental changes, and assessing existing and expected trends to predict impact.

Critics suggest that although the tool is an effective way of identifying issues relating to key elements in a macroenvironment, limitations exist because the results can be undermined by a fast pace of change or a failure of the analysis to identify complex interrelated factors (Evans et al., 2003).

Porter's Five Forces Framework

Michael Porter (1980) developed a model that considers competitive forces that can be used for destination planning. His analysis led to the construction of a framework that he proposed could assist a business in developing a competitive strategy by reflecting on five competitive forces, namely:

- Threat of new entrants
- Threat of substitute products
- Power of buyers or customers
- Power of suppliers
- Rivalry among businesses

This microenvironmental approach provides an interesting mode of analysis that cruise operators can consider when planning itineraries in a competitive environment.

Porter's Diamond Analysis and Associated Work

Developed in the 1980s, the Diamond analysis has in some respects a more direct application to studying ports or destinations because it was originally intended to be a study of regional or national competitiveness (Porter, 1990). This model identifies four factors that help to define a destination's competitiveness:

- Factor conditions—physical resources, human resources, capital resources, infrastructure, and knowledge resources
- Market structures, organizations, and strategy
- Demand conditions
- Related and supporting industries

Wahab and Cooper (2001) comment on developments to this work undertaken by Smeral (1996) to create a set of guidelines. These guidelines reframe Porter's original four factors to incorporate factor conditions within those aspects that are considered most relevant for tourism, namely:

- Market structures, organizations, and strategies—dealing with image and market position, product development and promotion, and desire for growth.
- Demand conditions—availability and development of quality facilities and services, seasonal influences, focus on tourist spending power, efforts to attract repeat visits, and presence of an integrated policy for tourism.
- Government—research and awareness of tourism market trends, availability of focused training, minimal bureaucracy, inclusive environmental awareness, and openness to proactive management of change.

Successful application of these analytical tools relies on the quality and reliability of contemporary data as well as a careful consideration of the interrelated nature of the stated factors.

Boston Consultancy Group (BCG) Matrix

This method of analysis was designed to consider a product portfolio and to make logical calculations about development (Knowles et al., 2004b). From a strategic perspective this approach can be applied to a cruise destination to consider growth rate, growth potential, passenger popularity, barriers for development and sophistication of facilities (Figure 5.2).

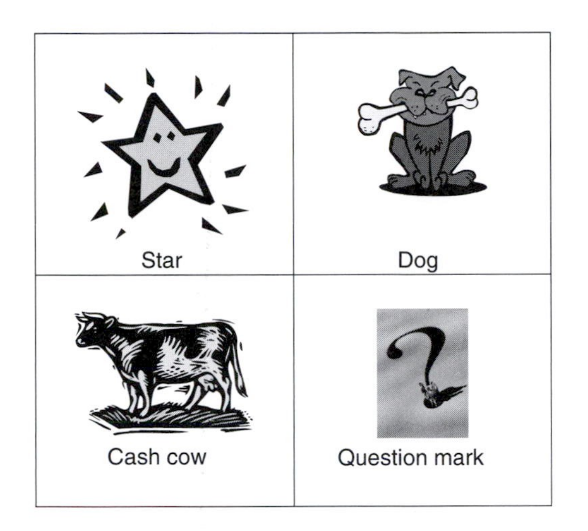

Figure 5.2: BCG matrix

TALC—Tourist Area Life Cycle

This theory (Butler, 1980) is predicated on the premise that resorts are products that have a series of evolutionary stages in relation to consumer demand (Knowles et al., 2004b). These stages are:

- Exploration: The resort is visited by a small number of tourists, access may be difficult.
- Involvement: Tourist numbers grow, with the result that basic services evolve and interaction increases with the local community.
- Development: The resort is promoted, external forces take an interest in gaining control of the resort, growth of visitor numbers accelerates to create an imbalance of locals to tourists at peak times, facilities are improved.
- Consolidation: Tourism is a major contributor to local economy, growth has leveled off, the resort has achieved an international reputation.
- Stagnation: The resort loses fashionable status, there is overreliance on repeat trade, various social and environmental problems arise.

There are two alternative directions that can emerge as the next evolutionary stage:

- Rejuvenation: There is a major impetus to reposition the resort through marketing and investment in facilities through a partnership of public and private sectors.
- Decline: The resort has a dependency on day rather than staying visitors and the use of visitor accommodation changes.

This theory presents interesting ideas for analyzing a resort, but there is another dimension to be examined in terms of the role cruise tourism plays in the life cycle of a resort or destination.

Doxey's Irridex

The term *irridex* is a derivation of the word irritation, thus explaining that this model considers Doxey's (1975) work examining potentially diminishing patience and support from the local community when faced with certain patterns of tourism. Shaw and Williams (2002) suggest that this model provides a useful lens through which to consider the impacts on a host community by reflecting on the community's reaction to tourism as an evolving industry.

In many respects Doxey's irridex shares Butler's (1980) overview of life cycle stages. The irridex charts progression from euphoria to apathy, then irritation, eventually through to antagonism, while Butler's model describes exploration, involvement, development, consolidation, stagnation, and decline or rejuvenation.

Components of the Destination Amalgam

Cooper, Fletcher, Gilbert, and Wanhill (1995) identified four "A's" that represent the components of a destination. In this model, each "A" is described as follows:

Attractions—These are described as being man-made (such as the Sydney Opera House or the Pyramids of Giza) or natural features (for example, the Grand Canyon or the glaciers of Alaska) or events (an Open Golf tournament or the Olympic Games). It is feasible for certain attractions to incorporate some or all of these characteristics.

Access—This component includes an analysis of transport links that considers roads, rail, airport, shipping access in the port or harbor, and the logical integration of these options to serve the visitor. Cooper et al. (1995:85) believe that the way a destination deals with supporting and providing "innovative provision" is important. This can include establishing walker's trails, cycling routes, horse and coach options, and helicopter rides.

Amenities—The line between a support facility and an attraction can be difficult to locate. Cooper et al. (1995) note that the attraction does what it says—it attracts—while the amenity is there to support the attraction. Therefore, restaurants, shops, bars, and hotels are generally considered amenities. For the cruise passenger this can be a moot point, because shopping complexes, high image hotels such as Raffles in Singapore, or a prestigious restaurant run by a famous chef can be attractors in their own right. That may well be the case, but this component also includes the availability of important localized services, such as toilets, swimming pools, entertainment, hairdressing, exchange offices, security services, and casinos.

Ancillary services—This component describes agencies or bodies that act to coordinate, develop, and market the destination, such as tourist offices, city tourist departments, and travel and convention centers. These services help the destination define a recognizable profile and image. The body or agency can provide leadership or facilitate cohesion for the various stakeholders in the destination and provide information, services, and advice to tourists or tour groups.

VICE

The term *VICE* is associated with environmental and sustainable practices for tourism (BTA, 2001). The acronym stands for the key guiding principles of Visitor satisfaction; Industry profitability; Community acceptance; and Environmental protection. This approach has developed potency in recent years as tourist authorities seek to establish strategies for sustainable development. As such it can be a useful measure for a cruise operator to identify when considering a destination and reflecting on the policy and practice that it adopts for sustainable development.

In this respect, it is also possible to consider other analytical approaches that can be undertaken for the destination that tie in to sustainability, such as heritage analysis, which addresses the notions of preservation, conservation, and exploitation (Smith, 2003). Another approach examines possible relationships between tourism and cultural heritage assets (McKercher and du Cros, 2002:16). It is possible to reflect on the various forms of analysis that are represented above and to construct through research a series of discrete tourist resources that typify a destination's attributes, such as natural attractions, human-made attractions, shopping experience, hospitality resources, restaurants, unpolluted environment, airport, weather, security, taxi and local transport, and friendliness, and then to rate them in order to create a cumulative and comparative score. In most cases, this approach is taken by cruise companies when they survey passengers. However, as Lockyer (2005) describes, this method of analysis can result in flawed understanding. While this approach is manageable and relatively easy to undertake, the bluntness of responses hide the subtlety of reality. This reality suggests that consumers make judgments based on complex factors that can be highly subjective because of individual circumstances. In this sense it appears logical to reflect on the consumers or passengers before making interpretations about the destination.

This final section on destination analysis reflects on ways that tourist motivation and drive can be studied to help planners understand the characteristics that make a destination attractive or not.

Abraham Maslow (1908–1970) is a central theorist when considering motivation (Maslow, 1970). His five-level hierarchy of needs model suggests that human beings are motivated by unsatisfied needs and that certain lower needs need to be satisfied before higher needs are. His model commenced with basic physiological needs (food and drink), then safety and security, followed by love or friendship, esteem and, finally, self-actualization.

Tourist Motivation

Cohen (1979) established a tourist classification that identified the following typologies:

- Organized mass tourist (package oriented—little contact with local culture)
- Individual mass tourist (as above but with personal choice)
- Explorer (comfort combined with independence—a unique experience with a safety net)
- Drifter (immerses in local culture—no itinerary)
- Institutionalized tourism (high on familiarity)

Cohen's typology was developed by Plog (1987) who was instrumental in creating a psychographic classification for tourists that is regarded by many as a seminal study. In his study Plog (1987) theorized that tourists were positioned on a continuum. Psychocentrics, who are "self-inhibited, nervous, and lacking desire for adventure" are at one end, with characteristics such as fear of flying, territorial boundedness, and general anxiety. The Midcentric occupy the central point, and the Allocentric, who are "outgoing and independent, keen to explore" are at the other extreme. Plog believed that the majority of the population could be found in the center of the continuum.

On the subject of desire and ambition for travel, Dann (1981) described seven elements of motivation, which are as follows:

- Travel is a response to what is lacking yet desired
- Destination pull exists in response to motivational push
- Motivation can exist as fantasy
- Motivation can be described as a classified purpose
- Motivational typologies can be described such as sunlust, wanderlust, etc.
- Motivation is affected by tourist experiences
- Motivation is a kind of auto-definition and meaning

This work can be compared to McIntosh and Goeldner's (1986) four categories of motivation: physical motivators, relating to body and mind, such as reducing tension; cultural motivators, such as the desire to see and know more about other cultures; interpersonal motivators, such as the need to meet new people; and status and prestige motivators, such as the desire for recognition and attention and personal development, to satisfy the ego.

Framke (2002) believes that decision making for tourists takes place at home; the push factors are more powerful than the pull factors. Framke asserts that in this sense the benefit of studying destinations and doing other tourism research is to help construct marketing plans that can attract the tourist to the destination. In cruising terms, however, there are more complex forces at work.

Logistics, Positioning, and Planning

Deciding upon an itinerary is a matter of identifying ports of call that meet customer needs as described earlier in this chapter: to be safe or nonthreatening, to be accessible, to be interesting, to be culturally stimulating and different from the everyday, to be friendly, and to be user friendly. From the perspective of the cruise operator, an itinerary aims to achieve a range of practical and logistical goals to ensure that the quality of the cruise experience is maintained when the cruise customer ventures ashore.

There are many examples of cruises that break the mold in terms of designing and planning an itinerary, yet the majority of companies fit the following pattern:

- Itineraries commence from a port of embarkation and conclude at a port of disembarkation, which may or may not be the same place.
- Many cruises are scheduled for 7, 10, or 14 days to correspond with customer availability and to meet customer expectation for duration.
- Some cruise operators schedule cruises to be cyclical (continually repeating an itinerary for a set period of time) or bi-cyclical (alternating between two co-located schedules over a period of time).
- Despite their size, many of the larger vessels are designed to hold only stores sufficient to comply with these standard cruising patterns.
- Many itineraries are aimed to create an arrival time at port in the morning and a departure time in the mid- to late-afternoon.
- Itineraries may make use of ports where ships can receive fuel called "bunkers", supplies, and stores (including food, drink, and drinking water—referred to as potable water); offload waste (from compactors and rubbish collection); and access specialist (technical) support services.
- Most itineraries maximize the number of days at port and minimize the number of days at sea.
- Cruise companies examine port costs carefully when selecting ports of call to ensure that the cost-benefit ratio is acceptable.
- Arrival and departure ports are selected with due regard for infrastructure in terms of onward travel, security, and terminal facilities and procedures.

In addition, a number of cruise companies aim to take advantage of their distinctiveness in scheduling. Factors that can set these companies apart include the following:

- Schedules to include world cruising (circumnavigating the globe), unique cruising (a different itinerary every cruise), short break or "taster" cruises, a cruise without a destination (to enjoy the ship as a destination), fly-cruising (attracting customers to fly to the embarkation port), sector cruising (a cruise that may be constructed from within a world cruise or lengthier voyage).
- Selecting exotic ports, less-frequented ports, and ports with more complex arrival and departure issues.
- Ports that provide connection and location for cruise and tour vacations.

In all these cases, critical itinerary-planning issues must be borne in mind by the operator. Among the most complex of these issues is the matter of border and passport control, immigration, and documentation, which will be considered in more depth later in the chapter. Passenger and crew health needs are also serious matters for consideration. In some countries travelers are more susceptible to illnesses caused by bacterial agents, mosquitoes or other insects, food contaminated by poor hygiene, and contamination of the local water supply. Preparation can help in some of these cases; for example, immunization can help by protecting the traveler from certain diseases, but there is also a need to inform customers and crew about risks so they can take appropriate action. Cruise companies take care to forewarn their passengers about potential risks and advise them to contact their doctor or seek appropriate advice if they need further information. In general, where it is deemed unsafe to drink water ashore, passengers should be advised to buy bottled water and avoid ice cubes and food items that may have been washed before consumption.

When the ship cannot safely tie up alongside the quay or jetty because of shallow water, the general condition of the approaches to the port in question, or the tidal variations, the itinerary may declare that the port is a "boat port." In these cases, passengers will land by way of the ship's launches or tenders while the vessel anchors safely offshore. Invariably, boat ports reduce the amount of time ashore because of the travel time to and from shore and the problems that may arise in meeting passengers' requests to disembark immediately. Boat ports require cruise personnel to establish and marshal a control point onshore to manage the process of arrival and return.

For vessels that return to ports on a regular or frequent basis, much can be gained from the familiarity of key personnel with port officials, agents and contractors. Continued experience with immigration helps the purser's department to smooth the process of arrival, disembarkation, transit of passengers (departing the vessel with the aim of traveling to another country), re-embarkation, and departure. The cruise company makes use of a port agent to act on the cruise ship's behalf as the shore-based facilitator and deal with a range of matters including official and immigration requirements, supply, onward travel, shore excursions, technical support, and specialized services.

Familiarity with the port is also important when advising passengers about local conditions and what to expect when they proceed ashore, and, overall, it helps cruise staff ensure that passengers have an enjoyable and safe visit. In some ports, the point of arrival is a considerable distance from the center of the town or the main point of interest, and the cruise company may need to contract a shuttle bus service to transport passengers promptly and safely to the desired location, where they can be left to their own devices to walk or find private transfers. Usually, shore excursion coaches will be at the ship's side or close to the passenger arrival point.

Planning the cruise is, as has been seen earlier, a matter of ensuring that the itinerary appropriately meets the needs of the target market while addressing a broad range of other internal and external factors. The next section considers these internal and external factors to further develop an understanding of planning issues.

The Elements of Planning

As Moutinho (2000) states, planning for tourism should be "integral"; that is, it should take a multidimensional and systematic approach so as to be viable in the long term. Operating an international cruise corporation does not allow for avoiding social responsibility, and, indeed, it would be counterproductive, in these days of rapid communication, for any business to be seen to be taking an ethically unsound approach to operations.

The global picture is complex in this respect because a cruise operator must understand the implications of visiting certain countries from a political, environmental, social, technological, legal, and economic standpoint. Each country is likely to possess a policy for tourism that will affect incoming tour operators, including cruise companies (Goeldner and Brent Ritchie, 2003). This policy, these authors suggest, should emerge from planning that takes a balanced view of economic value and social well-being.

Good tourism planning can provide long-term benefits to the local population in terms of the resulting tourist infrastructure (developing essential services, establishing effective transport, creating communication networks, and commercial facilities). The superstructure for tourism (Goeldner and Brent Ritchie, 2003), such as hotels, restaurants, car rentals, and attractions, while emerging in relation to tourists' needs, can also provide benefits for the local population.

Integrated planning linked with policy formulation ensures that destination management is strategically considered to maximize the benefits of tourism while mitigating the disadvantages that can emerge from tourism. Laws (1997), affirms that the "packaging" of tourism has resulted in four major outcomes: resorts emerge in response to demand as a type of homogenized replication of a standard model; environment and ecology are put under pressure by developments; the destination is presented selectively and in an oversimplified format that can, in turn, modify the way locals behave; and, while employment and commercial opportunities grow, there is a consequential penalty on the demand for infrastructure.

Cruise companies operate within this milieu. Their potential to affect destinations by disgorging an additional 2,000–3,000 passengers from each megacruise ship is great, and tourism planners must consider these factors in order to manage their effect on the destination sustainably. Equally, each cruise operator should aim to understand the destination in order to comply with regulations and local laws and to make sure the quality of the passenger experience is maximized.

Regulations

Tourism policy can affect cruise operators in a number of ways. In the first instance, complex regulations must be considered. These regulations may relate to a broad range of factors, including the mobility of people, goods and capital; health and safety laws; environmental protection; consumer protection; shipping; ownership of key facilities; and security.

In terms of the mobility of people, goods, and capital (Shaw and Williams, 2004), cruise companies may face the prospect of dealing with border controls. This can create issues relating to passport and visa controls, customs, financial exchange, and the passage of cruise passengers as they arrive at and depart from destinations.

A ship arriving in a foreign port must be cleared for arrival before passengers can disembark. This routine differs depending on the regulations that apply and the nationality of the ship by registration. The clearance is likely to involve the port authority receiving a declaration about the passengers, crew, and goods on board and information relating to the ship's itinerary. Goods and passengers joining or leaving the vessel while in port are also noted as part of this routine. In some countries, port health officials inspect the vessel to measure the levels of sanitation and hygiene on board. Increasingly, declarations are sought that confirm there are no health problems on board or that the ship presents no security risks (International Council of Cruise Lines, 2004b).

Recent problems regarding health exemplified by SARS and the norovirus, as well as terrorist threats, have changed the way that cruise ships and tourists travelling internationally are managed by port authorities. Heightened states of alert result in raised levels of security. Security regimes for cruise ships in US ports include the following security measures:

> 100% screening of all passenger baggage, carry-on luggage
> Intensified screening of passenger lists and passenger identification
> Restricted access to any sensitive vessel or terminal areas
> Stringent measures to deter unauthorized entry and illegal activity
> Notice given to U.S. Coast Guard 96 hours before entering U.S. ports, and passenger and crew identification information submitted to federal agencies
> Coast Guard-established security zone around cruise ships.
>
> Reference: International Council of Cruise Lines (ICCL) (2004b).

In Europe, an agreement among 15 of the member states called the "Schengen Treaty" was introduced in 1985 to facilitate the free passage of people between member states or treaty members (Eurovisa, 2001). The impact on cruise ships depends on the passengers' nationalities because, in theory, passengers from European countries that are part of the Schengen Treaty do not need to be processed through a border control. Complications can arise because of visa requirements, which may be different for non-Schengen and Schengen countries if, for example, a non-EU crew member has to be repatriated at short notice.

According to Eurovisa (2001), the name "Schengen" originates from a small town in Luxembourg. In June 1985, seven European countries signed a treaty to end internal border checkpoints and controls. As time has passed, more countries have joined the treaty, and currently there are 15 Schengen countries, all of which are in Europe: Austria, Belgium, Denmark, Finland, France, Germany, Iceland, Italy, Greece, Luxembourg, the Netherlands, Norway, Portugal, Spain, and Sweden. All these countries, except Norway and Iceland, are European Union members.

Regulations concerning port health can, as was stated earlier, result in a visit from health officials who will inspect any part of the ship to ensure that the vessel is operating safely and hygienically. The galley is often a primary focus because of the risks related to storing and preparing foods for consumption. Most large vessels employ an environmental safety officer who is responsible for ensuring that the ship complies with regulations and meets minimum standards.

Marketing and Demand

The aforementioned section identifies a feature that is stressed by the cruise industry: that it is the safest way to travel (International Council of Cruise Lines, 2004b). According to Goeldner and Brent Ritchie (2003), this gets to the heart of an individual's psychology or motivation: "a person is thus possessed of two very strong drives—safety and exploration—and he or she needs to reduce this conflict". The ICCL recognizes this conflict, but, in highlighting the industry's safety record and describing the ship as "comparable to a secure building with a 24-hour security guard" (International Council of Cruise Lines, 2004b), they aim to show how the regulatory framework actually enhances the passenger's potential enjoyment of their vacation.

More and more, cruise companies are adopting what is referred to as a "psychographic" approach to market segmentation (Goeldner and Brent Ritchie, 2003). Market segmentation is undertaken to divide a defined population into specific characteristics and thus maximize the potential of a marketing campaign. Traditionally, segmentation considered geographic factors (where individuals lived), demographic factors (their age, gender, family circumstances), and socioeconomic factors (occupation, social class, and income), but psychographic factors consider values, motivations, and personal issues. Thus, a cruise vacation can be sold as a "lifestyle" choice. From a planning point of view the implications are as follows: do the destinations and the products and services on board meet the psychographic needs and travel preference of the passengers?

Logistics

Logistical planning can take a number of guises. This type of planning can focus on supplies and services (fuel, provisions, or consumables), schedule planning (coupled with fuel consumption), or capacity management (maximizing efficiency when dealing with large numbers of people). Cruise ships can travel at speeds up to 25 knots. At this speed, the ship consumes fuel at a greater rate, although greater distances can be travelled. Cruise itineraries are planned so that the ship can travel comfortably between ports to ensure that:

- Fuel consumption is at an economically optimized rate
- The arrival time and departure time is as per the schedule
- The mix of destinations is appropriately balanced to meet customer needs
- Regulations are complied with

Cruise companies tend to include four to five ports of call for a 7-day itinerary and eight to ten ports of call for a 14-day itinerary (Laws, 1997). Increasingly, as the industry expands, there is a need to locate embarkation ports to ensure access to new markets (Goeldner and Brent Ritchie, 2003).

Shore Excursions

Cruise companies offer shore excursions or tours for a variety of reasons. Obviously, these activities generate revenue and make a vital contribution to a cruise company's bottom line, but the provision of tours also adds to the complete package that is the cruise vacation. For many passengers, the tour continues to develop the pattern of providing a secure and a hassle-free quality vacation. The tour is a relatively safe, easily organized and managed foray into a different culture or, alternatively, an opportunity to sample an activity. Through chaperoning or mentoring, this tourism experience cossets the individual.

Shore excursions are entirely optional and, therefore, sales and marketing are important aspects of managing this element of revenue generation. Sales are encouraged and undertaken even before the passenger joins the ship. In most cases, passengers or potential passengers are introduced to shore excursions in the cruise brochure, where summaries of key tours are included as a kind of appetizer. Tours identify the best options for spending time ashore and provide a vital link between the cruise ship and the destination. The choice faced by passengers is whether to be independent on arrival at the port or to leave the organization and subsequent implementation of the shore experience to the cruise company.

It is in the best financial interest of the cruise company to sell tours and yet, as is the case with marketing in general for cruise passengers, considerable care is required to ensure that sales are made with sensitivity to the setting. High-pressure sales techniques are counterproductive in this type of community setting where passengers are socially attuned to each other and experiences are shared. Shore excursion sales are more likely to be successful using a subtle selling approach (see Table 5.2).

Promoting Sales

The brochure is the initial marketing communication used to introduce the notion of shore excursions, and it is interesting to examine just how that is done. Emphasis is placed in the itinerary on key features and activities that can be experienced ashore. Often, these component parts are highlighted within specific tours. Some cruise brochures send a message to the cautious traveler that tours are the best way to take the risk out of visiting ports of call. Shore excursions receive prime billing in among options ashore, complete with clear guidance on how to book before joining the ship. Usually, the message contains a compelling prompt to avoid disappointment by booking the tour as early as possible.

When the tickets are sent to the passenger, the information packet that accompanies the ticket includes a shore excursion brochure. This is the primary vehicle for promoting sales before departure and prompting sales on board after departure. This reference material uses language most carefully in order to maintain an intellectual bond with the reader, to develop the appropriate style of communication, to be truthful, and to avoid contravening legislation related to misselling.

Virtually all companies use their websites to present their products to all who visit them online. A web presence is maintained as a marketing initiative and as a form of distribution. Prospective and existing clients can visit the website to find out more, to make comparisons, to make bookings, to communicate with the cruise company, and (in some cases) to communicate with other passengers. Provided the market has access to it, the cruise company website is an excellent vehicle for promoting shore excursions.

The itinerary and the operational aspects, such as deadlines for closing sales in order to confirm arrangements, can affect sales on board. Advertisements in the ship's newspaper, on the ship's television program (if available), promotions near the tour office, direct sales, and connections made by port lecturers all form part of the marketing plan. Just as important, a successful first port of call can generate additional sales via word of mouth.

It is also interesting to consider the timing of tours. Depending on the itinerary and port of call, ships may arrive at a port at 0800 and depart at 1700, which creates an opportunity to sell half- and whole-day tours. Many passengers realize that the tour presents the best, most time-efficient way to visit and experience the port of call. The added security factor is that most prestigious cruise companies guarantee to look after their passengers if a tour is delayed by delaying sailing or, if that is not an option, by ensuring they meet the ship at the next possible port of call.

Drivers for Sales

- Scarcity value
- Security aspect
- Best choice promotion
- Natural choice option

Table 5.2: **Sales options**

Cruise Brochure	Website	Promotions	Word of mouth
Tour Brochure	Ship's TV	Direct Sales	Port Lecturer

What Makes a Good Shore Excursion?

The answer depends on the passenger and the cruise company. As portrayed in the cruise brochure, ports of call or destinations are selected for a variety of reasons, one of the most important being the attractiveness of the destination to the passenger. Tours are offered by cruise companies to capitalize on those facets of travel that are most attractive to passengers. Here, passenger demographics play a part: tours are designed according to the type of passenger. In this way, families, older passengers, active couples, young singles, and all other identifiable market sectors can be satisfied.

The cruise company seeks a product that enhances the image of the cruise and the cruise brand. The tour must provide an itinerary that fits the ship's timetable. The logistics determined by the tour operator at the port must be in synergy with the requirements of the ship and passengers. At the same time, the tour operator must also assure health, safety, and security.

A cruise company's expertise or experience is likely to have an impact. Prior experiences are important in building relations with tour operators, in understanding what to expect on the quayside when dispatching tours, and in terms of recognizing which tours are popular and achieve best scores in passenger surveys.

Designing the Tour

The quality of communication between the cruise company, its agent, and the tour operator in the port of call is paramount. The port agent acts as a facilitator, although for many companies the communication between cruise company and tour operator can be direct and continuous. Initially, the cruise company constructs an itinerary taking into account logistical factors such as fuel and travel time, creating a balanced cruise program that will sell. The travel operator works with the shore excursions department to devise the shore excursion program, construct tour brochures, and plan resources.

The program will take into account the number and type of passengers on board the ship as well as the time in port and the availability and quality of transport such as coaches, sea or river craft, trains and helicopters or light aircraft. The availability of trained guides is also important. In many ports the expansion of cruising as a vacation has created the potential for increased traffic. As a result, several ships may visit ports on the same day, thus diluting the availability of shore excursion resources and increasing crowding. Arrival time at a port may be critical in gaining access to resources in a way that avoids overcrowding. However, cruise companies are less likely to include a port in an itinerary if overcrowding and quality control are concerns.

Most tour brochures include reliable favorites that are either a half day or a full day in duration. These include the "banker" sites—the primary reason that an attraction is an attraction. It is hard to imagine passengers not visiting the Pyramids in Cairo if the ship calls at Port Said or Alexandria. Full-day tours usually include a lunch. Morning tours are generally more popular than afternoon tours because passengers who book a morning tour frequently prefer to spend their time independently in the afternoon. Some cruise companies provide cars or minibuses (see Figure 5.3) to cater to individuals, couples, or smaller groups who prefer to remain apart from other cruise passengers. "Party" tours including music, dancing, drinks and food are popular with a younger demographic on island destinations. Cultural tours are popular with passengers from cruise ships where the emphasis is on discovery and learning. It also follows that cruises with themes generate interest in specific types of tours: for example, a vineyard visit in New Zealand may suit an individual attracted to a cruise with a gastronomy theme and a tour to the opera in Italy might meet the needs of a music lover.

Before Arrival

Tours are made available with minimum and maximum numbers. On leaving port, the shore excursion office would be aware of the actual numbers sold and remaining availability. The shore excursion team

Figure 5.3: Tour bus returning to the ship

would be briefed on the content of tours, the tours that are likely to be oversubscribed, the best alternatives, any special features, and tours that might suit specific types of passengers, such as those with walking difficulties. Many passengers like to talk to a member of the shore excursion team to get a feel for the tour, and for that reason it is helpful for the team to experience as many tours as they can so that they can share their knowledge.

As mentioned previously, the port lecturer is an important part of the sales equation. Many port lecturers coordinate with the shore excursions team to help inform the passengers about the contents of tours and to help passengers make the best choice for their enjoyment. Port lecturers also accompany tours as escorts and provide an additional point of quality control, feeding back to the shore excursion manager their own impressions as well as those of the passengers.

Approximately one day prior to arrival, the shore excursion manager will contact the tour operator with final numbers. The tour operator will already be apprised of the numbers before the ship sails and will have a general idea of actual resource needs. The tour operator can advise the ship if an extra allowance of numbers is possible to increase last-minute sales. The shore excursion manager can decide whether a tour that is close to minimum numbers should run or be cancelled and the tour operator compensated accordingly. This decision is often a matter of applying a cost-benefit analysis; for example, does the benefit of operating the tour and losing some revenue while satisfying customers outweigh the cost of cancelling a tour and creating dissatisfaction? In the situation where the shore excursion is not cancelled despite not achieving minimum sales, the anticipated loss might be cancelled by any last minute sales.

Sales ashore and on the ship generate a ticket and receipt for the passenger. In some cases, tickets can be produced that identify the passenger by name, providing extra security in case the ticket is mislaid. Advice regarding special conditions such as dietary needs, any requirement to wear walking shoes, and any need to be sensitive to dress appropriately for certain religious buildings is clearly communicated at the time of sale.

On Arrival

The imminent arrival of a ship to port triggers a series of actions that are intended to clear the ship formally for arrival by satisfying the port authorities that due process has been followed and all administrative tasks that were requested have been completed. When the port authority clears the ship for arrival, the ship can tie up alongside the quay or lie at anchor off the port as directed.

It is customary for shore excursion officers to be among the first people to go ashore to meet the tour operator, to check that arrangements are in order and that passengers can join their respective tours (see Figure 5.4). The process of disembarkation requires careful planning to ensure that the correct passenger gets to the meeting point for the correct tour at the correct time.

Passengers are usually asked to meet at a gathering point close to where they will disembark so they can receive an adhesive color-coded badge. The ship's staff communicates by radio to coordinate disembarkation and guide passengers to the relevant tour point. Many cruise companies encourage ship's staff to accompany tours as escorts. This enables the ship to have a representative with the passengers who can, if necessary, act for the company and also provide comment about the quality of the tour after the passengers get back on board. Escorts are generally fully briefed by the shore excursion team and are provided with a checklist to complete. Tours are marked simply and visibly with codes that correspond to the badges worn by passengers.

A shore excursions officer, who registers the return of passengers to the ship, meets returning tours. The tour operator and the shore excursion manager agree on the numbers of passengers who have taken tours so that the cruise company can make payment to the tour operator.

Other Duties

In addition to selling tours, the shore excursion office is a tourist information office and a travel agency. In its capacity as a tour information point, the staff is frequently asked to supply information

Figure 5.4: Joining the tour

about ports of call, including basic information about the distance from ship to town or more complex information relating to custom and practice. Most shore excursion offices hold data files to help staff answer these questions and also rely heavily on staff to develop their own knowledge with experience.

As a travel agency, staff may be called on to arrange hotels or book onward travel arrangements such as flight tickets, train tickets, and organizing taxis or ferries. This side of the business also generates revenue through commission on sales. Some cruise companies operate a separate travel company ashore that can deal with these passenger needs on demand.

Tour Guides

Good tours rely on good tour guides (see Figure 5.5). The interaction between a tour guide and passengers is essential to the success of a shore excursion or tour (Collins, 2000). This individual entertains, informs, and organizes to varying degrees depending on the needs of the passenger. As the scale of the cruise industry continues to expand, the need for high-quality tours and guides follows suit.

In general, this type of person has an in-depth knowledge of his or her field and is highly skilled as a communicator. This aptitude can include fluency in languages, a well-tuned but carefully practiced sense of humor, and an ability to empathize with a broad range of people. Frequently, a guide will also have the ability to perform first aid and be able to assert her or himself whenever necessary.

Many tourist organizations operate an accreditation scheme for tour guides to ensure they are appropriately qualified. Frequently, it appears, guides are over-qualified, possessing higher-level degrees in their subject and being regarded almost as experts in their subjects. Guides can also benefit from being more mature in order to have had the breadth of life experiences that can lend themselves to this type of job. In addition, all guides must be physically fit because of the rigors of the job.

Figure 5.5: Tour groups in Kusadasi

Table 5.3: **Points for good practice: the tour guide**

A good practice guide:

1. Ensure you have all contact details and emergency numbers (just in case).
2. Establish a good rapport with the coach driver, go over the itinerary for the tour and identify potential problems that may not have been foreseen (such as roadworks).
3. Make sure you know the details relating to the itinerary including method of payments for access to sites etc.
4. Test the microphone—ensure it is in good working order and can be heard in all parts of the coach.
5. Practice using the microphone; many people hold the microphone loosely against the chin because it is consistently at the correct distance from your mouth in this position.
6. Make sure you get a courier's seat or front seat; check the seat belt.
7. Inspect the inside and outside of the coach for cleanliness and general condition. This is the responsibility of the coach driver but you should reassure yourself that the coach is in appropriate condition (no cracked windows, tires in good condition, clean, no damaged panels, lights and air conditioning working etc.).
8. Greet clients; smile!
9. Before departure, do a head count.
10. Before engine starts, introduce yourself and point out safety aspects as directed.
11. Sit before commencing commentary.
12. Check that your clients are OK and are listening—glance back.
13. Don't talk too much and adopt an appropriate pace and intonation of speech.
14. Start by saying what the tour is going to be with an outline of stops, comfort breaks and meal breaks.
15. Be precise about return to coach times; repeat these messages to stress timings and help passengers to identify the coach by appearance, number and location.
16. Be particularly clear if coach might have to relocate after stopping and passengers debus.
17. At meals best to sit with the driver unless an alternative arrangement has been made.
18. Passengers are likely to expect toilet stops and souvenir stops.
19. When directing, say things like "on the right you will see" and always make sure you don't talk over a particularly interesting sight.
20. Don't forget camera or photographic moments.
21. Be identifiable—umbrella, your clothes, a hat—something that is easily remembered and easily seen.
22. Be vigilant about hazards. Uneven pavements, low ceilings etc.
23. Know your passengers: find out a bit about them and use that in your talk (if appropriate that is).
24. Keep language simple and be willing to answer questions. Remember the folk at the back may not hear the question so repeat it to them before giving the answer.
25. Include all age groups by aiming to provide a commentary for everyone that might be on the tour.
26. Explain local rules and customs so that people know why things are as they are.
27. Keep counting and checking at key points (off and on the coach).
28. Keep your group together and keep pedestrian traffic flows clear.
29. Be proactive to try and avoid problems.
30. Use positive body language and maintain eye contact.
31. Develop a strong routine at the end to mark the final point off the tour.

Attractions, tour operators, coach companies and other tourist venues can employ guides. Some are freelance or self-employed. Guides are used on walking tours, coach tours, within notable buildings or sites (such as art galleries, cathedrals, or castles), as trail guides, as sports guides, or as interpreters. For many cruises, there may be some tours that are particularly popular, and the guide may have to plan with the tour manager how best to orchestrate visits and to maximize passenger enjoyment.

Tour guides may work with tour managers, who coordinate the shore excursion provision, and coach drivers. In addition, they tend to develop good working relations with those employed by the attractions or key locations that they are visiting. The guide may also need to identify any tour escort (normally an employee of the cruise company). See Table 5.3.

Summary and Conclusion

This chapter examines the very nature of what it takes for a destination to be an appealing port of call. Thereafter, a number of theoretical approaches to analyzing and evaluating destinations and tourist motivation are suggested. These approaches should be considered carefully to make sure that the appropriate theory meets the task at hand. To make such a judgment, the reader is advised to further review the literature to examine the theories in more depth. The chapter concludes by describing shore excursions operated by cruise companies. This section of the chapter includes a critique of tour planning and tour management.

Glossary

Berth: The part of a harbor or dock area where a ship can be accommodated and passengers can disembark and embark.
Circumnavigate: To sail completely around the world.
Demography: The statistical study of human populations, especially with reference to size and density, distribution, and vital statistics.
Generating region: The location where cruise passenger business is sourced.
Indigenous population: Originating from the area.
Product portfolio: A variety of goods and services that together form the range of items to be sold to customers.
Stakeholders: Groups or individuals who have an interest in the venture.
Strategic: To achieve an overall plan.
Tactic: To achieve a goal.

Chapter Review Questions

1. What is a destination?
2. How can destinations be analyzed?
3. What makes a good cruise destination?
4. What is Schengen?
5. How are shore excursions organized for passengers on a megacruise ship? Consider:
    a. sales and marketing
    b. organization and planning
    c. arrival at port and joining the tour
    d. evaluation and quality control

Additional Sources of Information

Cruise companies and shore excursions
http://www.cunard.co.uk/Destinations/default.asp?Region=12&sub=se
http://www.princess.com/planner/shorex/index.jsp
http://www.pocruises.com/shoreEx/home.htm
https://shorex.rccl.com/
http://www.crystalcruises.com/onboard.aspx?ID=9#id44
http://www.ncl.com/shorex/index.htm
http://shorex.celebrity.com/cc/home.asp

Third-party sources
http://www.shoretrips.com/default.asp
http://www.cruisedirect.com/act_shoremain.shtml
http://www.cruisecritic.com/tips/tipsarticle.cfm?ID=25

Port consortia or general information
http://www.medcruise.com/page.asp?p=1516&l=1

http://www.port-of-call.com/Ports_of_call_v2/default.asp
http://www.soulofamerica.com/cruises/american_ports.html
http://www.econres.com/documents/Cruise.html
http://goflorida.about.com/od/cruiseports/

References

BTA (2001), The Sustainable Growth of Tourism to Britain. London: British Tourist Authority.

Burton, R. (1995), *Travel geography* (2nd ed.). London: Pitman.

Butler, R. (1980), The concept of a tourist area cycle of evolution, *Canadian Geographer* (24).

Cohen, E. (1979c), Rethinking the Sociology of Tourism, *Annals of Tourism Research*, 6(1), 18–35.

Collins, V. R. (2000), *Becoming a tour guide*. London: Continuum.

Cooper, C., Fletcher, J., Gilbert, D., and Wanhill, S. (1995), *Tourism principles and practices*. Harrow: Longman.

Dann, G. (1981), Tourist Motivation: An Appraisal, *Annals of Tourism Research*, 8, 187–219.

Davidson, R., and Maitland, R. (1997), *Tourism destinations*. London: Hodder and Stoughton.

Doxey, G. V. (1975), A causation theory of visitor-resident irritants: Methodology and research inferences, Paper presented at the Travel and Tourism Research Associations Sixth Annual Conference Proceedings, San Diego.

Eurovisa (2001), Free movement of people. Retrieved 11 October, 2004, from http://www.eurovisa.info/SchengenCountries.htm

Evans, N., Campbell, D., and Stonehouse, G. (2003), *Strategic management for travel and tourism*. Oxford: Butterworth Heinemann.

Framke, W. (2002), The destination as a concept, *Scandinavian Journal of Hospitality and Tourism*, 2(2), 92–108.

Gibson, P. (2002), The GNVQ in leisure and tourism: investigating student perspectives. In B. Vokonic and N. Cavlek (Eds.), Rethinking of education and training for tourism. Zagreb: Graduate School of Economics and Business, University of Zagreb, Croatia.

Goeldner, C., and Brent Ritchie, J. R. (2003), *Tourism: principles, practices and philosophies*. New Jersey: John Wiley and Sons.

International Council of Cruise Lines (2004), The safest way to travel: cruise ship security. Retrieved 11 October 2004 from www.iccl.org

Knowles, T., Diamantis, D., and El-Mourhabi, J. B. (2004), *The globalisation of tourism and hospitality: A strategic perspective* (2nd ed.). London: Thomson.

Laws, E. (1997), *Managing Packaged Tourism*. London: International Thomson Business Press.

Lloyd's. (2003a, April/May), 'Cruise International.'

Lloyd's. (2003b, January), 'Silverseas.' Lloyd's Cruise International.

Lockyer, T. (2005), Understanding the hotel accommodation purchase decision, Paper presented at the CHME Conference, Bournemouth.

Maslow, A. (1970), *Motivation and personality* (2nd ed.). New York: Harper & Row.

McIntosh, R., and Goeldner, C. (1986), *Tourism principles, practices, philosophies*. New York: Wiley.

McKercher, R., and du Cros, H. (2002), *Cultural tourism: The partnership between tourism and cultural heritage management*. New York: Hayworth Hospitality Press.

Melian-Gonzalez, A., and Garcia-Falcon, J. M. (2003), Competitive potential of tourism in destinations, *Annals of Tourism Research*, 30(3), 720–740.

Moutinho, L. (Ed.), (2000), *Strategic management in tourism*. Wallingford: CABI Publishing.

Peattie, K., and Moutinho, L. (2000), The marketing environment for travel and tourism. In L. Moutinho (Ed.), *Strategic management in tourism*. Wallingford: CABI.

Plog, S. C. (1987), Understanding Psychographics in Tourism Research. In J. R. B. Ritchie and C. Goeldner (Eds.), *Travel tourism and hospitality research* (pp. 203–214). New York: Wiley.

Porter, M. E. (1980), *Competitive strategy: Techniques for analyzing industries and competitors*. New York: Free Press.

Porter, M. E. (1990), *The competitive advantage of nations*. New York: Free Press.
Shaw, G., and Williams, A. M. (2004), *Tourism and tourism spaces*. London: Sage.
Smeral, E. (1996), Globalisation and changes in the competitiveness of tourism destinations in Peter Keller (Ed.), *Globalisation and Tourism, 46th AIEST Conference,* Rotorua, New Zealand. St. Gallen, Switzerland.
Smith, M. K. (2003), *Issues in cultural tourism studies*. London: Routledge.
Wahab, S., & Cooper, C. (2001), Tourism globalisation and the competitive advantage of nations. In S. Wahab and C. Cooper (Eds.), *Tourism in the age of globalisation*. London: Routledge.

6

Working on Board

Learning Objectives

By the end of the chapter the reader should be able to:

- Understand the roles and responsibilities on a cruise ship, the personnel structures, contracts, and organizations relevant to cruise vessels
- Identify the role and importance of the purser's department and hotel services for the cruise and shipping industry
- Reflect on the shipboard culture and the challenges of managing a multicultural crew

This chapter examines staffing and personnel organizational structures and the concomitant roles and responsibilities on board cruise ships. The role of the purser is considered and the hotel services department described to help the reader understand the lines of authority and importance of human resource relationships on board. Crewing is a complex issue for cruise operators, and particular attention is given to the mechanisms that exist to support the management of hotel services crew, staff development, and the sourcing of skilled, semi-skilled, and unskilled labor for the hotel services department. The organizations that exist to provide assistance and support to international crews, often in defense of the crew member's basic employment needs, are identified together with a discussion relating to this sensitive employment issue (Garrison, 2005).

Describing the purser's role and the on-board hotel services creates a link to introducing the range of products and services provided to customers and crew. Doing so, in turn, introduces the distinct areas of operation that are defined as hotel services, including customer services, bars, lounges, restaurants, bistros, and sleeping accommodations. In addition, those services that are concerned with customer entertainment, such as shops, casinos, entertainment cast, and port lecturers, are discussed.

Cruise ships are likely to be heterogeneous; that is, containing a mixture of crew with different nationalities, of various ages, with different backgrounds and prior learning, and individual needs and aspirations. The latter part of this chapter asks questions about managing such a multicultural and diverse crew and provides case studies that are intended to highlight crew perceptions about life on board.

The Role and Responsibilities on a Cruise Ship

Most large cruise ships boast extensive facilities and activities, which necessitates employing a virtual army of people to ensure the "resort" operates to meet guests' needs. Traditionally, ships employed officers and ratings (non-officers) or crew who performed tasks related to the safe passage and commercial activity of the vessel within a hierarchical regime. This regime was often operated on a rotating "watch keeping" basis, from which the term "officer of the watch" is derived. Automation has, on many vessels, changed the strict pattern of 24-hour watch keeping, but for any ship there remains a need to maintain operational effectiveness, safety, and security. Watch duties are traditionally 4 hours in duration: 0800 to 1200, 1200 to 1600, 1600 to 2000, 2000 to 2400, 2400 to 0400 and 0400 to 0800. Typically, a deck or engineering officer will undertake two 4-hour watches in a 24-hour period.

On cruise ships, the same hierarchical regime exists, for reasons that are explained later, but, unlike a tanker or cargo vessel, the majority of employees are associated with customer services. In contemporary cruising, employees are designated officers, crew, and staff (Bow, 2002). Officers are employees with specific authority. They are located within four departments: deck, engineering, radio, and hotel services. The crew is similarly divided among these four departments, but in number this group represents the largest segment. The last group, "staff," includes personnel such as shop managers, hairdressers, beauticians, entertainers, casino staff, and photographers many of whom may be contracted to work on board by a concessionaire. The organizational chart in Figure 6.1 is an example of structure for managing a large cruise ship.

The resulting "ship's company" is a large and diverse community that, because of scale and complexity, requires careful management and coordination. A ship's master has, according to Branch (1996), absolute authority aboard a cruise ship. This authority, acting in lieu of the ship's owner, provides powers to act accordingly in cases where the ship, crew, customers, or ship's contents are at risk. This responsibility is further outlined in the discussion of maritime law in Chapter 3. There is a subtle difference between this role and the post of captain, which is deemed to be a rank, although frequently the captain on a cruise ship will hold the position of master. Other officers on board may also hold the rank of captain, such as the staff captain who is charged with a responsibility relating to the crew, staff, and customers.

The Deck Department

The ship's master is in charge of the ship but also oversees navigation and the deck department. On a day-to-day basis, the deck department is the responsibility of the chief officer or first mate (first officer). The larger the vessel, the greater the requirement for additional deck officers, who are termed second, third, or fourth officer (the exact number depends on the size of the ship). This department oversees navigation and the care of the vessel. One of the senior officers in this department will also hold the position of safety officer. The deck officer's complement is frequently made up of junior officers-in-training, who are called *cadets*.

Crew positions include chief petty officer (deck) and petty officers (deck), who supervise deck crew under the direction of deck officers; the deck carpenter, who attends berthing and departure; the quartermaster or coxswain (senior rating), who is a responsible for steering; junior

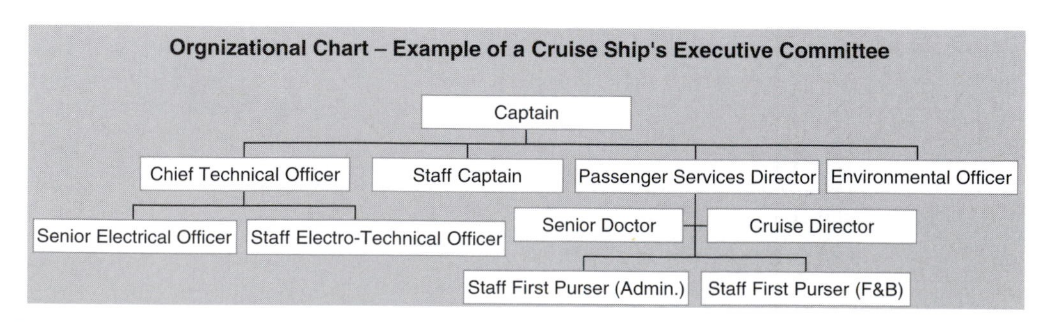

Figure 6.1: Organizational chart – management structure

seamen, seamen grade 2, and seamen grade 1 (the latter two share lookout and steering duties with deck officers); the Bosun (boatswain), who is the deck hands' overall foreman; and day workers employed in general duties. All seamen employed in the deck department who are not officers fall into the category of ordinary seamen (OS), who are deemed to be unskilled, or able-bodied seamen (AB), who are considered skilled. The deck department can also include specific posts, such as security.

Deck officers can be identified by their stripes, which are plain gold. Masters and captains have four stripes, chief officers have three stripes, first officers have two-and-a-half stripes, second officers have two stripes, third officers have one-and-a-half stripes, and fourth officers have one stripe. Cadets frequently have either half or one stripe. The symbol for the deck department is a diamond. Security can be recognized by their brown stripes and their symbol, the capital "S." Most cruise ships appoint an environmental officer. This person is answerable to the ship's captain and is recognizable because he or she wears green and gold stripes.

The Engine Department

The engine room is the domain of the chief engineer, who is responsible to the master for the vessel's propulsion, steering, and power for auxiliary systems such as heating, ventilation, air conditioning, lighting, and refrigeration. The chief engineer is also responsible for fuel, maintenance, and repairs. Depending on both the size of the ship and the type of propulsion system, cruise ships may require additional engineering officers and cadets, including electrical engineers.

Crew positions in the engine department include the chief petty officer (motorman) and petty officer (motorman), who supervise the engine room under the direction of engineer officers; junior motorman, motorman grade 2, and motorman grade 1. Some vessels have specific posts, such as electrician.

Chief engineers have four stripes, which are alternately gold and purple. The chief electrician has three stripes, the first engineer has two-and-a-half stripes, and the second engineer has two stripes. There are two symbols for this department; the propeller signifies technical and engineering, while an electric current motif is used for electrical officers.

The Radio Officers

There is usually a chief radio officer, who may be supplemented by additional radio officers depending on requirements on board. The radio office supports all communication, including radio, telex, telegraph, telephone, and Internet and satellite communication. Radio officers and radio assistants (who are ratings employed to assist the officers) are affiliated to technical and engineering but they work very closely with the navigators, and, for that reason, the radio room is generally located close to the ship's bridge.

The chief radio (or communication) officer has three stripes, which are gold and green. Other radio officers follow the same pattern as deck and engineer officers. The symbol for the radio or communication department is a radio signal. This post is fast disappearing. The world of electronics means that the technical role is more likely to be managed by an electrical officer.

The Medical Department

It is not surprising that, given the size of the community on board, a cruise ship requires a medical team. The principal medical officer (PMO), supported by as many medical officers, or doctors, as are required, leads this department. Depending on the ship and the clientele, there may also be a senior nurse or two or more nurses (usually at officer level). Some vessels also employ orderlies, who tend to be designated as ratings. The very largest of ships may also employ a medical dispenser, physiotherapists, and dentists. Some ships have a morgue on board.

The medical officer is usually identified as having three stripes, which are gold and red. The symbol for this department is the caduceus (staff of Hermes). The provision of medical support on board is a necessity for the well-being of the shipboard community. The medical team can also generate revenue in providing specialist support, and for that reason some cruise companies locate the medical team under the management of the hotel services department.

The Entertainment Department

The cruise director, who is usually an experienced professional from the world of entertainment, leads this department. Any aspect of entertaining customers (and crew) is managed from within this department. The range of employees can therefore include musicians, dancers, comedians, actors, singers, social hosts, sound and lighting crew, stage technicians, guest lecturers, port lecturers, health, fitness and sport instructors, children's staff, and specialist experts.

A deputy cruise director frequently assists the cruise director. The cruise director is usually regarded as having a rank equivalent of three stripes and is linked by association to the hotel services department.

The Hotel Department

Depending on the scale and size of operations, the hotel service team usually dominates in terms of numbers of employees. An individual with the title of hotel manager, director of hotel services, passenger services director (PSD), or executive purser is usually in charge of the department. The term *purser* is, traditionally, related to the controller of finances, hence, the derivative of "purse". However, different cruise companies use the term in different ways. The senior officer in charge of hotel services has four stripes, which are gold and white. The executive chef, the food and beverage manager and the deputy purser have three stripes. The senior assistant purser, assistant food and beverage manager, bar manager, and accommodation manager (housekeeper) have two-and-a-half stripes. The second purser has two stripes. The symbol for this department is the cloverleaf.

Depending on the particular cruise company, the focus on core values, the type of passengers, and the product on offer, the hotel services department may be configured to reflect more traditional nautical lines or may represent more contemporary hotel services as seen shoreside. Because this department is the focus of this book, the various roles are outlined in more depth in the following section.

The Management of Hotel Services

Everything to do with managing hotel services on a cruise ship reflects the scale of the vessel and the labor intensity associated with product and service quality. In addition, if the ship and the tradition related to the development of cruising as a nautical enterprise are deemed to be a valuable marketing focus, the hotel services team is likely to be led by the executive purser. Otherwise, the senior role might be identified as the passenger services director or hotel director. Hotel services are line managed by two or three senior managers who may have the job title of deputy purser. These deputies focus on food and drink, passenger services (including accommodations), and finance. In turn, each deputy will lead a team that may comprise the following:

Food and drink: Executive chef and kitchen brigade, including bellbox chefs; bar manager, bar supervisors, bar stewards, assistant bar stewards; maître d'hôtel, restaurant managers, head waiters, head sommelier, assistant sommeliers, waiters and assistant waiters; crew and officer's mess chefs; and stewards and utility stewards.

Passenger services: Accommodation manager, accommodation administration, accommodation, supervisors, public area supervisor (decks), public area supervisor (lounges), utility stewards, cabin stewards, butlers, laundry master, assistant laundry master, and laundry assistants.

Administration and personnel: Administration manager/assistant purser front office, assistant administration manager, junior assistant pursers/front office manager, receptionists, crew assistant purser, shore excursion manager, shore excursion assistant purser and junior assistant pursers.

Additional areas: Shops, florist, print shop, administration stores, art auction, communication center, beauty center. Photography may also be located within the remit of the post holder.

Various cruise companies operate hotel services to suit their strategic needs. For example, Princess Cruises recruits a team of junior assistant pursers who are delegated to specific responsibilities, including reception, shore excursions, art auction, food and beverage, and the crew office. They report to assistant pursers who are section assistant managers on board. In some cases, when the operation requires, the assistant pursers report to senior assistant pursers.

The provision and orchestration of food and drink and accommodation are demanding from a human resource (HR) perspective. The effective performance of any cruise ship is underpinned by the quality of service provided by people such as waiters, accommodation stewards, sommeliers, and public service stewards. This presents a serious challenge to cruise companies. In periods of growth, cruise companies are faced with seemingly mind-boggling HR requirements. New megacruisers require in the region of 1,000 new personnel. To staff a ship, contractual arrangements and the patterns of contracts must be taken into consideration. These staff require training, supervision, and management. Using experienced personnel to carry out training duties or to establish new vessels into service dilutes the skills base on what may be a cruise ship with an impeccable reputation. The challenge to achieve minimum standards is neverending. An example of an organizational chart for the purser's department can be seen in Figure 6.2. Note that many companies operate their hotel departments using alternative job titles.

In the galley, the team is frequently configured using a variant of more traditional approaches to the "hotelerie"-style brigade of chefs. This revolves around the executive chef, supported by a team of sous chefs who control the hot plate (referred to as the "hot press" on some vessels) where service takes place in the galley. These sous chefs may also be needed for service to satellite restaurants, depending on the style of production and expected standards of food service. The various sections in the galley are managed by *chefs de partie*, such as larder chef, butcher, sauce chef, grill chef, fish chef, and pastry chef, who work with their assistants. In addition, there are breakfast chefs and bellbox

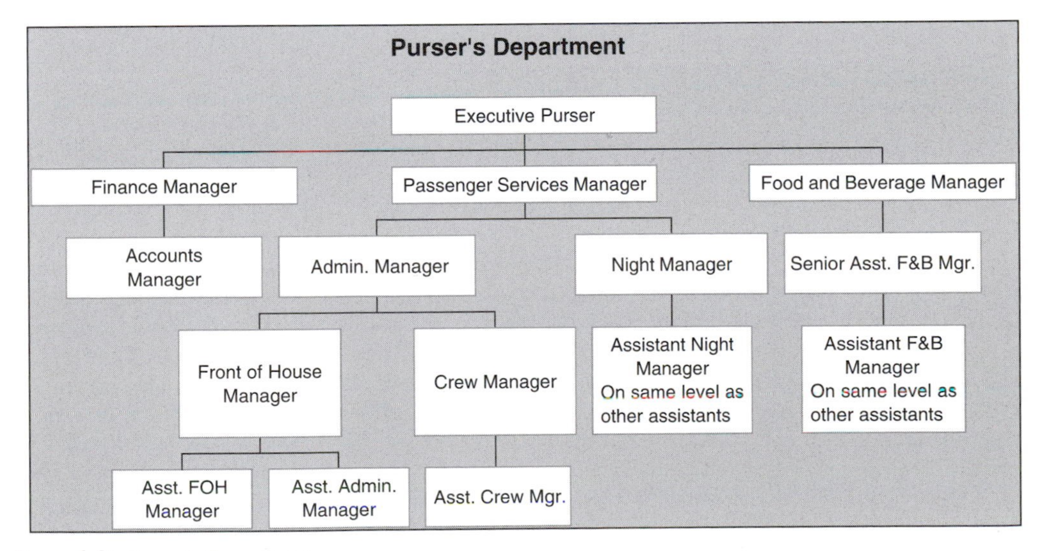

Figure 6.2: Purser's Department

chefs. The latter produce food for room service from a pantry galley close to cabins or staterooms. Finally, it is very important to recognize that the officers, staff, and crew also need to be fed and it is the job of the chefs who manage the crew and officers' messes (the name given to the ship's personnel dining areas on board) to manage this complex and demanding task.

The restaurants on larger ships operate using a supervising maître d'hôtel (maître d'), who coordinates a team of restaurant managers. Each restaurant employs a host to welcome guests and facilitate their entry and seating in the restaurant. Head waiters are allocated to sections of a large restaurant (often up to 300 covers per section) where they have oversight of waiters who work with an assistant or "busboy" to serve 16 to 18 covers. These types of large restaurants frequently employ a head sommelier and a team of assistant sommeliers to serve drinks and wine. Most megacruisers include a buffet service area, as well as fast food outlets such as pizza bars or burger bars. These are operated using a team of assistant buffet stewards supervised by head waiters or assistant head waiters.

The bar manager on a vessel organizes a multibar operation that can include a dispense bar for restaurant drink sales, show bars, cocktail bars, lounge bars, deck and pool bars, champagne and caviar bars, crew bars, and, in some cases, an officers' mess bar. The bars are coordinated by bar supervisors who work with the individual bar stewards and their assistants.

Food and beverage managers working with assistant managers/pursers coordinate operations with the various managers. Stores managers receive, store, and issue goods to the relevant personnel working in galley, bars, restaurants, or point of requisition.

The Shipboard Culture: Managing a Multicultural Crew

The crew aboard a megacruiser is likely to be diverse by nationality and culture. For a vessel with approximately 2,000 passengers, there may be 900 to 1,000 crew members. On an early summer cruise in the Mediterranean the makeup of a Princess Cruise ship's company was as follows:

Number of passengers—2,054
Number of crew—980
Nationalities of crew—54
Nationalities of passengers—64

The following case studies present a synopsis of what it is like to work within this type of community. Names are disguised to create anonymity.

Case Study: Juanita, Junior Assistant Purser

Juanita is Mexican. She came to work for Princess Cruises after studying business administration and tourism. She had an idea that she would progress to work in a hotel but became aware of the opportunities working at sea from a relative. She had always dreamed of traveling and, in particular, harbored a desire to go to Italy. She remembered that she used to keep travel brochures so she could imagine different places around the world and she even had a separate collection of Princess Cruise brochures. Initially, she started as an assistant buffet steward but eventually transferred to the purser's desk, where she is employed as a junior assistant purser. She really enjoys her life at sea, which she feels is good experience for a single female, and she has ambition to progress in management on board. One aspect of her work that she enjoys is the job rotation, which means she could be working in different areas, such as shore excursions or the crew office, on different contracts.

Juanita is fluent in English and in Spanish, her mother tongue. She notes that there is a diversity of nationalities and various first languages on board and states that it is amazing how people get along. She is friendly with people of all nationalities and declares that one critical factor for this international situation is that the brain doesn't have a language and the crew is all working together. Rank may well create some separation but, in general, people get along really well.

Work on the reception desk is demanding and challenging. Passengers went there if anything had gone wrong or to find out information. Juanita believes that reception staff need to be both strong and caring about people, with a "How can I help?" orientation. Problems have to be managed and passengers supported, and while the outcomes are usually positive, the journey can be difficult. Juanita gets time off in port because the office team is large enough to allow for its members to cover each other and maintain appropriate staffing levels. However, it is not always possible to go ashore on every occasion. Juanita has come to appreciate that it is good to learn languages and to be people oriented. This is the type of job where the people you work with are also the people you socialize with. This invariably means that you soon become friends and at the end of the contract you have an added bonus in being able to visit them all around the world. Juanita has been all over the world while working with Princess Cruises, and recently she went to Italy for the first time.

Case Study: John, Staff Purser

From a career that commenced with hotels in the UK, John moved to cruise ships to work with a company that had formed when P&O cruise merged with Princess Cruises. He had the choice to join either P&O Cruises out of Southampton or Princess Cruises. He chose the latter because he found the US style of service appealing. In his current position he is a senior manager with an ambition to progress to the post of passenger service director (PSD) in the near future. He started work in reception as an assistant purser (AP) and was then promoted regularly until he achieved his current position. John now covers for the PSD. Until five years ago, John believed that the company was very slow to promote on ability. The situation changed, and promotion is now based on merit, informed by a sophisticated appraisal system. This change meant the company had a large group of people in place who were very keen and extremely talented. In the shore-based office there were people who were used to the former and more traditional P&O Cruises purser model; P&O Cruises has since modified their approach to managing this department. John states that most managers are highly experienced in people management. He declares that responsibilities are demanding and salaries are appropriately rewarding. Accordingly, managers are unwilling to allow for any interference from secondary sources that can impact on the effectiveness and quality of operations.

Crew members receive gratuities based on a point system, with 65 percent going to the food and beverage department and 35 percent going to the accommodation department (there is a per-day charge added to the passenger account to cover gratuities). Gratuities are awarded up to the accommodation manager and the maître d'. The base salary is less than it had been previously, but John commented that the gratuity made up for this change. While passengers think this situation is not motivating, John says that, in reality, it is. With a 15 percent service charge added to drink sales, there is a visible impact on up-selling and it is also an inducement for staff to seek more training. Princess Cruises operates on a credo known by the acronym CRUISE, which stands for Customer, Respect, Unfailing In Service Excellence. This was the eighth year it had been operating, and all staff appeared to be aware of it. Passenger satisfaction has leaped. It used to be 82 percent but now the company is unhappy if the satisfaction level is less than 90 percent.

In John's opinion, the current cruising season for the ship is going well, which can be attributed to many things, including the itinerary and an absence of unforeseen problems such as an outbreak of Norwalk virus (norovirus) or engine problems. The focus for managers on board is predicated on the need for awareness. It is better to predict and prevent problems rather than be faced with the repercussions. His job gives him a wide remit of responsibilities, including hotel key system (security), appraisals, line managing senior assistant managers, gift salon, art manager, on-board sales, administration, hotel stores, florist, linen, photography, upholsterer, carpet, control of stores, review of purchase orders (not shops), standards of performance, accommodation, compliance with regulations, indents, having the ship in readiness for dry dock, the logistics of embarkation and disembarkation, Pratika (dealing with port officials), security situation, tendering operations, the on-board newspaper (*Princess Patter*), overall printing, crew office, berthing, and ship's crew welfare.

Cruise ship managers may well have more autonomy than was the case in the past, but the vessels are not isolated islands operating independently from their corporate home base. Technology allows shore-based managers to be aware of circumstances and events onboard, so that critical issues identified ashore can be addressed with greater immediacy. This mode of response has come about, in part, because of the tendency for some to regard litigation as the most appropriate form of gaining a response to emerging problems. This response is also warranted because ships are subject to audits and inspection, which are in the public domain and can affect reputation.

John is a member of a number of committees. The executive committee (formerly the PSD meeting), the captain's conference, and the task force meetings are all revenue related. In addition, there are meetings of the Norwalk virus action committee, the purser's office meetings (SAP), the junior assistant purser meetings, hotel stores monthly meetings (which focus on costs and deal with any situation that could lead to excessive waste) and the cruise meeting (which involves all departments and used to be known as the ships committee meeting). The cruise ship is a lively community. People who are good at their job are noticed and rewards can be extremely good.

These cases draw attention to the demands of working in the hotel services department and highlight the difference in jobs and responsibility between a junior assistant purser and staff first purser. Elements that are worth considering further include those aspects that define the dynamics and atmosphere onboard, the attention to professionalism, the notion that this is a team environment, and the strength of opportunity for long-term careers.

Working on Board: Practical Considerations

New employees who join a cruise ship can be somewhat overwhelmed by the environment at first. The scale of the ship, the way of life, the structures that support and inform crew, and the disciplines of working at sea are all potentially alien and take some getting used to. Invariably, people work on cruise ships because they can travel, but there are other advantages as well given the types of jobs, levels of remuneration, opportunities for promotion, and conditions of employment. There is no getting away from the fact that this is a unique job that entails long hours and long contracts. Hospitality businesses often cater to people who are socializing or purchasing services outside their normal work times, and it is not new that some characterize hospitality jobs as involving work that takes place during unsociable hours. This view ignores the sociability of hospitality work and the benefits in working when many people are playing, and playing when those same people are working (Douglas and Douglas, 2004). The cruise setting provides a break from stereotypical work patterns that many find repetitive and tedious.

Conditions on board will depend on the employer, but the best provide excellent crew facilities and operate management regimes that are in tune with the personnel on board. A successful cruise ship resonates with the harmony created by personnel who are proud of what they are doing, who know they are good at what they do, and who enjoy what they do. Dining facilities revolve around a crew and officers' mess that is often serviced from the main galley but has a separate sous chef and team to provide meal that may be specifically designed to meet the cultural and religious needs of specific groups of crew. Catering for crew is a large-scale, 24-hour operation. It is important to feed the crew well so they are happy with this element of their lives on board. Some officers may be expected to eat with passengers as part of their public relations duties.

Going ashore depends on operational circumstances. Where the team can cover duties, a split can be planned to create some time off for all members of a particular group or department. If the facility is closed and the personnel have no additional duties, time off can be organized. For example, shop personnel can generally take time off to go ashore as all shops must close when the ship is in port. If duties are sufficiently demanding, personnel may not be able to go ashore. Under maritime regulations ships must maintain minimum staffing levels for safety and security reasons.

If a member of the ship's complement is ill, he or she will be expected to see the ship's doctor and will be treated appropriately. In a worst-case scenario, a crew member may be sent ashore to be treated, and possibly repatriated. In less serious cases, the person will return to work after recuperation. In most situations, health and safety regulations prevent a person who is unwell from handling food and drink and serving passengers.

Crew members are generally accommodated in serviced and shared cabins. The more senior personnel are allocated larger cabins, and over a certain rank single en suite cabins are provided. Space on board any cruise ship is limited, so crew members are advised to take care with what they bring on board. The standard and specifications of crew accommodation varies depending on the cruise company and the ship. As with the dining arrangements, it is in the interest of the cruise company to provide the best standard of accommodation possible to ensure that crew are satisfied with this element of their life on board.

Social life is generally a high point on board. Working and living in close proximity with colleagues invariably promotes a high level of camaraderie. Crew members can spend time in the crew bar, officers' wardroom, or social areas. Generally, prices of drinks are considerably lower than they would be in passenger areas. Most cruise ships have a crew club representative who organizes special events for the crew after consulting with the crew committee. The crew also has access to specific facilities that can include a pool, hot tub, gym, and cinema. The ship's captain and senior officers manage discipline. When a crew member joins the ship, she or he signs on to confirm that he or she will comply with the regulations. Serious breaches of discipline can mean that a crew member will be dismissed instantly.

Recruitment Practices

Finding a job on a cruise ship can take some investigation. Some companies act as a form of introductory agent, charging the applicant for the benefit of gaining access to potential employers. Others are genuine agents who are intermediaries in the recruitment process, often with offices located close to or within countries that are targets for employment. Some agents are actually secondary companies established by the cruise company to facilitate recruitment. Finally, some cruise companies employ directly. Trade journals, for either shipping and nautical matters or hospitality and catering, can be a useful source of information because major employers use them to gain access to a more specialized and experienced applicant. Managers in the purser's department should have an appropriate undergraduate or postgraduate higher education qualification, which may be based on business, hospitality, or tourism. Alternatively, many employers recognize professionals who have experience working and managing in the hospitality industry but who may not have formal qualifications.

The University of Plymouth in the UK operates a three-year undergraduate degree (four with optional work placement), the BSc (Hons) Cruise Operations Management, that can help those who are seeking a hotel management position aboard a cruise ship prepare and qualify for this type of work (University of Plymouth, 2005). Applicants should think more than twice before parting with money to secure an introduction or get help with finding any type of job on a cruise ship. The best starting place for many appropriately qualified applicants is with the cruise companies themselves.

Finally, applicants should remember that the work and lifestyle might not suit everybody. There are many examples of potential crew members who were inadequately prepared for their experiences on board and who had to leave at the first chance or be repatriated because they were unable to acclimatize or work their way through. It is not in the best interest of a cruise company to have to deal with serious human resource retention problems, because it is potentially disruptive for other crew members, can impact on service quality, and is costly from a selection, recruitment, and training point of view. Klein (2002) presents personnel problems on cruise ships in an interesting light. His study represents the negative position. Many of the points raised in Klein's book are typical of comments made in opposition to globalization and capitalism as vehicles for a fair world and are punctuated by his stance questioning the cruise industry as a socially and environmentally sustainable entity. He posits a view of an industry that, at times, suggests that employers are oppressive or exploitative. Klein's work is an interesting read. He constructs a set of arguments that should be read by those who are seeking to work in this industry so they can consider the criticism and reflect on a balanced view of life and work for contemporary cruise companies. Research is presented in a later chapter on training that provides a contradictory view to Klein's.

Summary and Conclusion

The environment aboard a cruise ship is a society in microcosm. The society is one with a clear purpose: the operation of a cruise. Yet, as the case in any society, there are subtleties and nuances pertaining to the diversity of individuals and the need to manage this environment to create equilibrium. The command structure described in this chapter creates a framework designed to sustain management of operations. From there, managers direct and support teams in a way that establishes the normative conditions for working on board. This chapter has highlighted some of these issues, which are examined further in Chapters 11 and 12.

Glossary

Bell box: A pantry for room service.
Butler: A hospitality professional who provides a high caliber of personalized service.
Hierarchical regime: A system that depends on rank and responsibility at different levels.
Multicultural: Coming from a variety of different cultural backgrounds.
Officers' mess: Officers' dining area.
Sommelier: A wine waiter.
Sous Chef: The second in command in the galley.

Chapter Review Questions

1. What are the major departments on a cruise ship?
2. What function does each department have?
3. What are the options available for employment?

Additional Reading and Sources of Further Information

http://www.british-shipping.org/training/index.htm: UK Chamber of Shipping
http://www.knowships.org/: US Chamber of Shipping
http://directory.fairplay.co.uk/showarea.asp?area=Ship+Owner: Directory

References

Bow, S. (2002), *Working on cruise ships*. Oxford: Vacation Work Publishing.

Branch, A. E. (1996), *Elements of shipping* (7th ed.). Cheltenham: Nelson Thornes.

Douglas, N., and Douglas, N. (2004), *The cruise experience: Global and regional issues in cruising*. Frenchs Forest, Australia: Pearson Education.

Garrison, L. (2005), Finding a job in the cruise industry. Retrieved 2 May 2005, from http://cruises.about.com/cs/cruisejobs/a/cruisejobs.htm

Klein, R. A. (2002), *Cruise ship blues*. Gabriola Island: New Society Publishers.

University of Plymouth (2005), BSc (Hons) Cruise Operations Management, from www.plymouth.ac.uk

7

Customer Service

Learning Objectives

By the end of the chapter the reader should be able to:

- Comprehend the importance of operations and management on customer service
- Describe the range of customer services
- Compare and contrast the internal and external factors that influence customer services
- Examine customer service systems for cruise lines
- Discuss profiles of cruise customers and specific needs
- Reflect on demographics and segmentation

Service and Quality

This chapter is concerned with customer service and the rather elusive issue of quality. The term quality presents a number of complexities, starting with the basic idea that it is a form of utopian excellence (Tse, 1996). Consider, for example, what quality means from the perspective of an operational manager aboard a cruise ship. Harris (1989) draws attention to the way that quality can correlate to prestige relating to reputation, admiration, luxury, and consequently, as a result, price. Compare Silverseas Cruises with Easy Cruises, for example, or consider that quality can be the way the customer views the service received against the perception of what was offered. For some, quality is concerned only with ultimate customer satisfaction.

Taken a stage further, some authors suggest that quality refers to a product or service that is a combination of the predictably uniform and the reliable, suitable for the market and at the lowest cost (Deming, 2000). Other theorists proclaim that it has more to do with the customer's perception of what is fit for the purpose (Juran, 1980). Yet others propose that quality has to do with creating "zero defects" and getting it right first time (Crosby, 1996).

This debate suggests that managers should pay some attention to clarifying, in their own minds, what they believe quality to be so that they can establish goals and targets. The implications for operations are critical to achieve the desired level of service together with the appropriate standard of product to budget. The subtleties of adopting an approach that strives for excellence at all times can create interesting operational dilemmas within a service organization that relies on staff and customer interaction. For example, cost control and consistency of practice are two such areas of concern. Compare this to an organization that understands what its customer wants and designs the service to meet or even exceed this goal (Harrington & Lenehan, 1998). Wright (2001:186) declares that, if an organization claims its service or product is of high quality, the implication is that, beyond the basic service or product specifications, there are so-called "higher level benefits" such as attention to detail, a high level of courtesy, and those little (or sometimes not so little) things that differentiate the business.

W. Edwards Deming, Philip Crosby, and Joseph Juran are key sources who were involved in promoting a managerial stance that focuses on quality. Their work, and the involvement of others, has led to the implementation of processes that are predicated upon the notions of researching quality, designing service quality as a fundamental for competitive differentiation, assessing perceptions of quality, and making improvements (Tse, 1996). Their particular views of quality in business embody subtle differences (see Table 7.1), yet they all share a common belief that managers can make the difference, that improvements can be achieved only by involving stakeholders (such as suppliers), that improving quality is not easy, and that the process is ongoing (Harrington and Lenehan, 1998).

According to Dale (1999), Total Quality Management (TQM), or Total Quality Control (TQC), was adopted in order to create continuous improvement, taking the lead of consultants from America, such as Deming, who contributed to the business renaissance in Japan during the 1950s and 1960s. Their approach to achieving continuous improvements was to identify a best practice, ensure it was established as best practice, and train the workers to achieve that best practice. Tse (1996) charts the progress of TQM as a management (and staff) philosophy within production-oriented companies through to its adoption by service companies. She lists the five guiding principles as: "commit to quality, focus on customer satisfaction, assess organizational culture, empower employees and teams, and measure quality efforts" (Tse, 1996:303). The implications are far reaching because this approach creates a cultural change and, for some companies, change can be too slow and results may not be instantanious (Wright, 2001). However, TQM can establish a particular approach for managers that is well suited to service organizations.

Proponents of TQM emphasize that change for all organizations is inevitable and striving for continuous improvement by embracing change is desirable. "Kaizen," a Japanese term for steadfast day-by-day betterment, is the name given to this process of continuous development. The word has entered the vocabulary of successful companies who have absorbed the kaizen approach into normal business practice (Wright, 2001).

Table 7.1: **Management of quality**

Quality theorists	*Crosby*	*Juran*	*Deming*
Key phrase	Zero defects. Right the first time.	Fit for purpose.	Predictable degree of uniformity and dependability at low cost and suited to the market.
Key focus	Quality is established if the product or service conforms closely to customer requirements.	Quality can be set by stating goals and then aiming to achieve these goals.	Quality should be consistent, reliable and acceptable. Quality is everyone's business.
Implication	Time and resources are not spent on correcting errors.	Teams, groups and individuals should be appropriately organized and trained.	Customers have precise and clear needs and will go elsewhere if these are not met.
Management action	Confirm who is the customer and how they define quality in the setting. Senior managers responsible for ensuring quality.	Managers would need to be clear about the purpose and what constitutes fitness for that purpose.	Managers should reduce delays, mistakes and defective work. Managers need to understand the market and how it is changing.

It is also useful to reflect on the causes of poor quality. Harris (1989) lists some of the common factors that have been linked to claims of poor quality:

- A lack of concern for quality within the organization
- Elements of the organization omitted from the quality drive
- Incomplete or unavailable specifications for products, services, and processes
- Badly designed operational methods
- Poor supervision and management leading to too much or too little discipline and control
- Poor supervision and management leading to poor morale
- Badly trained or untrained personnel
- Poor working conditions
- Poor job specifications
- Poorly maintained equipment and tools
- Materials not to purchase specification
- Ineffective audit or inspection
- Poor senior management control
- Lack of rewards and incentives

This list suggests that quality is highly complex and that circumstances where quality is at issue are highly individual. In addition, the variables connected to quality output are such that getting it right requires commitment, consistency, and professionalism.

Quality of Products and Services

Industry observers state that there are fundamental differences between products and services that are important when designing standards and establishing quality thresholds (Harrington and Lenehan, 1998). In essence, they can be described as follows (see Table 7.2).

As referenced earlier in this chapter, the key to ascertaining quality is to understand customer needs and wishes. However, at all times the entity seeking this understanding is the provider of the service, and this implies the potential for misinterpretation. The problem can arise because, over time, people or processes involved in a service can change, thus jeopardizing both the actual service quality and the understanding of customers needs. A TQM approach stresses the need to continually appraise both aspects and to invest in training so as to ensure the consistent and ongoing management of quality. In this way, a cruise company can focus on identifying what the customer wants and then aim to provide that within a defined budget.

Table 7.2: **Defining products and services**

Product	*Service*
A thing/object/device	A deed/performance/effort
Tangible	Intangible
Stand alone as an item	Requires people to take part
Customer not involved	Customer fully or partially involved
Standardized	Heterogeneous—different every time
Can be stored	Perishable
Can be tested prior to sale	Cannot be sampled prior to sale
Production often separated from consumption or usage	Production and consumption often occur simultaneously
Product purchase involves variable opportunity for reflection	Service encounter is a moment of truth

Operations and Management

The success of a cruise business, in terms of securing repeat customers and capturing new business, is directly related to reputation. Past and present customers and their perception of service and product quality directly inform that reputation. As discussed earlier, much is written about service quality and customer perceptions (Dale, 1999; Peters, 1987; Williams & Buswell, 2003; Wyckov, 1982) and the notion that to achieve quality, an organization must forever strive for continuous improvement. Ultimately, the customer defines the level of service that is appropriate.

It follows that cruise companies that focus their attention on meeting and, indeed, exceeding customer expectations of service and product quality will be in a stronger position to retain existing customers and attract future customers. Companies invest time and money in order to prioritize their customer service programs so that both staff and customers recognize the importance of getting customer service right. However, there may be a yawning chasm between promoting customer service initiatives and delivering effective customer service, and it is certainly not easy to deliver consistent, high-level, quality service. Disgruntled employees, an unexpected event such as an itinerary change, and production problems resulting in interruptions to service delivery are among potential threats to maintaining service quality.

The formula for a successful cruise is demanding. Getting everything right and exceeding expectations means ensuring that officers, managers, crew, and staff are trained, are instinctively customer oriented, are empowered to help customers if there is a problem, are aware of expected quality standards, and are capable of exceeding those standards. All this has to be done consistently and within budget. A crew member who has been on board for a nine-month contract has to be as fresh in his or her approach as an employee who is a new arrival.

Customer service presents serious challenges for managers at sea for a number of reasons. Staffing ratios of crew-to-customers can be high (almost 1:1 on luxury vessels), thus creating a requirement to ensure that all crew who are in contact with customers are suitably customer focused. Customers on large vessels may be diverse in terms of their countries of origin, which can mean that expectations of quality in customer service will vary. Contemporary cruise customers are demanding, in part because we live in a media-rich society that highlights consumer rights and advocates the benefits in complaining.

Nonetheless, it should also be recognized that being at sea has many customer service advantages. Staff are contracted to work on board for a number of cruises. In this situation, crew members cannot easily withdraw employment, nor can their performance be hidden from supervisors. The interaction aboard a cruise ship is complex. Customers and crew are together, forming relationships, for many days. In this situation, customer service is ongoing, and there are likely to be many occasions when crew members can provide service that can turn an ordinary vacation into something special.

One additional element is worth considering. There are varying levels of labor intensity required on cruise ships for different activities. Food service and food production are examples of high labor intensity, compared to the equivalent shore-based hotel model where ratios of staff-to-customers have been shrinking. In part, this is influenced by the desire to differentiate and to maintain "service quality," but cruise ship staff numbers are also affected by important regulatory and safety issues.

Operations management in hotel services aboard cruise ships is configured to maintain optimum service contact strategies to meet service quality parameters. Decisions are taken to ensure service contact is essential and to make use of alternative approaches when appropriate (Lovelock, 1992). For example, bookings for special services or shore excursions can be made prior to embarkation through the website or a travel agent, thus reducing the need for personal contact on board. This type of management action can reduce lines, or queuing; can reduce congestion, can increase customer satisfaction, and can increase staff job satisfaction.

In addition, consider events on a cruise ship that involve the potential for passenger interaction with crew and officers. These events may call for anything from high contact to low contact. Strategies can be considered to reduce contact in a way that either improves service quality or leaves it unaffected. For example, using on-screen account records in cabins reduces the numbers of passengers visiting

Table 7.3: **Contact strategies**

Contact reduction strategy
Aim to use the phone, mail or other form of contact for most contacts.
Introduce reservation and appointment systems.
Create secondary information points to take pressure away from the main facility.
Use drop-off points to collect customer information.
Bring services to customers.
Make use of roving greeters to control, entertain, and give information.
Use signs judiciously.
Contact improvement strategy
Use take-a-number system.
Train contact personnel to deal with all situations they are likely to meet.
Maintain consistent operating hours.
Partition the back office from the reception area.
Develop queuing or line patterns with signs.

the reception to collect and, in some cases, query invoices or folios. At other times, careful organization of events, such as the captain's cocktail party or welcome meeting, can create an impression of high-level contact without necessarily affecting operational demands. Lovelock (1992) believes that all service organizations can create operational improvements by reflecting on service contact in the following ways: decoupling services where it is of value to do so; aiming to reduce contact wherever possible by using appropriate strategies (see Table 7.3); where contact is inevitable, aiming to enhance the contact to benefit all parties; examining low-contact areas to continuously address efficiency and quality improvement.

The level and type of contact aboard a ship can vary. The range includes the personal service the passenger receives in the dining room, the interaction between cabin steward and passenger when they meet in the stateroom or cabin, passengers being joined by officers at their table in the restaurant (not practiced on all ships), and meeting officers in the bar and nightclub in the evening. Most passengers report that these forms of contact, and a broad range of others that are not described, are critical in maximizing the passenger experience (Douglas and Douglas, 2004). It is also important to recognize that crew, staff, and officers need to recharge and be able to spend time away from passengers in order to remain focused on the customer.

Managing Customer Services

The typical range of services that may be available on a typical Grand class cruise ship catering to a US market is presented in the order that might be experienced by a passenger on board (see Table 7.4). The list is not inclusive and is intended only as a guide. As cruising continues to grow, niche markets are targeted and innovative products and services are introduced accordingly, so this list is not exhaustive.

Information

In any vessel with a large number of customers on board, information must be communicated accurately, effectively, and in a timely fashion. Most cruise ships operate a pursers' office or a reception desk to provide a focal point for customers in need of information. Initially, when customers embark,

Table 7.4: **Services on board**

Embarkation	Butler service
Welcome aboard	Leisure services
Orientation and induction	Sport and recreation services
Safety and lifeboat drills	Beauty and health treatments
Food and drink service	Entertainments
Shops and boutiques	Casino
Medical service	Nightclub and disco
Port lecturer and information services	Shore excursion
Accommodation services	Disembarkation services

there is a settling-in period as they find their way about the ship. Information is provided to help with this task: Information is sent to the customer's home address, an information pack is placed in the staterooms or cabins, and printed information resources, such as the cruise "news", are placed in various key locations on the ship. Invariably, some customers will head for the reception desk to get answers, some may use the telephone, others will stop and direct questions at crew members, regardless of who they may be.

From a customer service perspective, there is much to be gained in predicting customer needs. While this approach is important throughout the cruise, there is evidence to suggest that first and last impressions are important in setting a template for service perceptions and sealing that set of perceptions about service experiences (Office of Quality Management, 2005). Equally, such predictions can help to establish a planned set of routines so that staffing levels at the reception desk, training of staff to deal with embarkation queries, and production of printed material can be coordinated to best effect. The negative effect of poor customer service in dealing with information can lead to dissatisfaction, congestions, and lines at information points such as the reception desk. It can also cause an overload of telephone inquiries, creating non-response or late response to queries, a semblance of ineptitude, or a lack of concern and professionalism.

Influences on Customer Service: Tipping

Whether employed as a waiter, a cabin steward, a public room steward, or a bartender, service staff who interact directly with customers are likely to receive gratuities, or tips. Throughout history, tipping has been a constant, yet sometimes awkward element of the guest–staff relationship. In European hotels and restaurants, a system was devised called the "tronc." That practice created a model for distribution of shared tips, which was replicated for all hospitality businesses. This system allocated points based on a hierarchical system based on rank in the service domain. Those with higher points, such as the maître d'hôtel (often referred to as the maître d') or the restaurant manager, received a larger slice of the total receipts. This system relied on those receiving the tip to submit the money to a central pot.

Cruise companies operate vastly different systems of tipping. In part, this reflects cultural differences from the passenger point of view, but there can also be other elements, such as the company or brand perspective. Thus, one company can proclaim that the ship is a no-tipping zone, as is the case with *Ocean Village*. Another company, such as Princess Cruises, levies a $10-per-day service charge to each passenger, which is then divided in tronc-like fashion. Passengers are also automatically charged 15% on every bar bill for the same reason. In both cases, the service charge is stated to be optional and that the passenger should act to remove the charge if it is deemed inappropriate. P&O Cruises provides prospective passengers with a guide to tipping, so as to encourage passengers to provide a gratuity to specifically identified staff based on a formulaic approach.

It is useful to consider the different stakeholders in a scenario involving tipping. Each provides a different perspective on the transaction and the implication that arises from the act.

Stakeholder 1—The Passenger

Different passengers react to tipping in different ways. If a customer is used to tipping it becomes almost second nature. If a customer is used to tipping as a general exit strategy, the notion of reward for extra special service becomes rather less of an issue. Perhaps at times, the tip is in recognition that the server is in some way underpaid, so it is to right a wrong. Some passengers tip as an entry strategy. It is a message to the server to be attentive and that there is a promise for more to come. Some passengers identify a key individual, such as a headwaiter, as the focus of this entry strategy. Passengers are generally happy to pay a set amount per day as a service charge and to pay a percentage on top of all bar bills, but some inevitably decline and act to remove the service charge because they think it is either unfair to them or it is not what they want to do.

Stakeholder 2—The Server

Money is money and, in a service job, tips can provide the bulk of income, making the job financially viable. At times, the tip can be received in a spirit of genuine reward, as a thank you for making the vacation special. Other times, some passengers hide the tip in an envelope to disguise the fact that there are only a couple of dollars. Tipping can be incredibly unfair. Sometimes it depends who the customer is and the good luck or bad luck of getting either "Ms. Generosity" or "Mr. Mean." Passengers do not seem to understand what tipping really means to the server. Some are uncomfortable and get embarrassed, as if it is a dirty act. But it does not embarrass the server—it is too important for that. A managed system using fixed daily payments is fine, especially for supervisors, and there is the added benefit that tips may still be given in addition to the levy.

Stakeholder 3—The Employer

Tipping is an essential component of the cruise experience. It allows employers to pay minimum rates with the understanding that actual income will be acceptable to the employee. If the money were not acceptable, the staff would not renew their contracts. That said, managed badly, tips can be a source of potential discord and disharmony. Staff notice what fellow employees receive and at times some servers can feel unhappy if they did not receive an anticipated tip. A managed tipping system using percentages and fixed payments gets around this problem, but it is noticeable that some passengers still tip anyway, which can mean the problem is diminished but not eliminated.

From the other angle, a tip-free environment is an interesting approach to managing the potential problem of passengers and crew reacting negatively to tipping. In stating that tips are not expected and are, in fact, discouraged, a message is sent by the operator that tipping can be in some way unfair, that some customers can be uncomfortable about tipping, and that they are above such petty matters. Furthermore, it appears the operator is suggesting that their staff members do not require to be tipped (presumably they are appropriately rewarded through their payment), that the staff is happy about this situation, and that good service is not to be bought as an extra—it is all part of the package.

The Human Side of Service Quality

Customer service can be affected by personal factors related to life on board. Cruise ships operate employment contracts for fixed terms. In some cases, the contracts can be 6, 8, or 10 months in duration. Crew members work every day for these contracts and are expected to be consistently effective. Some cruise companies have a reputation for operating more "enlightened" employment policies than their competitors. Those with better reputations ensure that there is a fair and open approach to time offshore, covering for illness or unavoidable absences from the work area, and to maintaining a quality social environment for all on board.

Customer service can be at risk if the server is unhappy for whatever reason and is in need of a break. The nature of a shipboard community is that it can be a happy and almost sheltered environment where the people who work supportively together are friends and companions. If there is a breakdown in that arrangement, it can be uncomfortable, and it is not in the interest of the cruise company for staff to be unhappy in either their social time or their work time.

Many cruise companies ensure that the crew elects or appoints a social club director to work with paid employees to construct a program of events and activities. Despite the apparent monotony of having to work lengthy contracts, reality for the crew is different. The attraction of travel and the places that the crew visits, coupled with the food package, inexpensive drink, entertainments, a lively social life and the use of phones and the Internet as a means of staying in touch, can mean that life on board is frequently more attractive than life at home. Many crew members have reported that, after a month at home, they were looking forward to returning to work.

Customer Service Systems for Cruise Companies

Customer service strategies are often adopted to orient the brand holistically and to clarify the brand's vision and mission to customers. Examples include the following:

CRUISE—The "Princess Cruises Program" (which stands for Courtesy, Respect, Unfailing In Service Excellence) recognizes and rewards shipboard personnel for developing solid customer relationships, for being proactive and responsive, and for going out of their way to meet our customers' needs with a smile (Princess Cruises, 2005). P&O Cruises and Swan Hellenica also use this acronym.

Gold Anchor Service—(Royal Caribbean Cruise Lines, 2005) sets forth the following philosophy:

> What is Gold Anchor Service? It's how we make your cruise adventure even more memorable. One thing that keeps our guests coming back again and again is our friendly and personal service. Maybe it's a server who remembers the name of your daughter's teddy bear. Or the bartender who remembers the extra olive. Or perhaps the housekeeper who reminds you of your dinner reservation time. Our unique style of service will enhance every aspect of your cruise. No matter where you are—the pool, the dining room, the spa, or your room—get ready to be wowed! And we deliver it 24 hours a day. This is way beyond normal service. This is Gold Anchor Service.

Likewise, Silverseas seeks to go above and beyond all expectations:

> Silverseas' service is simply the world's best. It is a philosophy, an attitude—complemented by distinctive European style and inherent in all that we do. Achieving perfection is driven by our desire to please, to see you smile. It begins the moment you step aboard with a warm welcome and a flute of champagne, and follows throughout your voyage with an unspoken anticipation of your needs. Sailing on Silverseas' intimate ships is like visiting a friend's home; you're greeted by name and your personal preferences are always remembered (Silverseas Cruises, 2005).

Demographics, Profiles of Cruise Customers, and Specific Needs

Cruise passengers are attracted by direct and targeted marketing. The product is designed specifically with certain people in mind and the cruise brands are very focused on the customers to whom they are selling. This strategy creates excellent levels of knowledge about who is likely to be on board, but within the typical profiles that emerge there is likely to be a broad range of specific needs. Most of these needs can be predicted and catered to, but there are always going to be unforeseen instances—individuals have new needs or requirements that add an extra challenge to operational management.

The cruising demographics change annually, with lower age groups and a broader range of customer types beginning to have greater impact. In the past, the stereotypical cruise passenger may have

Table 7.5: **Passenger needs**

Profile
Married couple in their 40s, both working, active lifestyles, enjoy finer things in life. Socially adept and aspirational, enjoy seeing new places and meeting new people.
Couple in late 20s with young family (son 6 and daughter 4). He is an Information Technology professional, she works part time from home as telephone researcher. Want to relax and keep kids happy.
Single female, retired, aged 72. Fit, healthy, and active. Likes dancing and meeting people. She is an experienced cruiser.

been of pensionable age and female. Older passengers, passengers with a disability, passengers with specific preferences (for food and drink, on-board entertainments, and specific ports of call and shore-side activities) were understood and their needs addressed as well as could be achieved. For example, boat ports, with a tender from ship to quayside, always presented a problem for those passengers with trouble walking or who travelled in a wheelchair. Changing demographics create new demands. For example, products and services offered in response to demographic shifts include gyms, children's nannies, computer games, action sports, more casual dining, and larger nightclubs.

Table 7.5, above, identifies the types of needs each passenger is likely to require.

According to the Cruise Line International Association (CLIA, 2005), there is a range of "Personas" or personalities, each with a particular set of psychological attributes (Williams, 2002). These persons are synonymous with the types of people who cruise. Such personalities include restless baby boomers, representing 33 percent of all passengers, who, it is suggested, may find cost to be an impediment to trying different vacations; enthusiastic baby boomers, representing the 20 percent of cruisers who are described as living a stressful life and seeking an opportunity to escape; luxury seekers, the 14 percent of all cruise passengers who unsurprisingly spend the most; consummate shoppers, the 16 percent of passengers who look for best value; the 11 percent of cruisers, described as "explorers," who are well educated; and the 6 percent who are "ship buffs," or seriously interested and knowledgeable about cruise ships. This last category is also the most senior segment.

Providing Customer Service

Customer service is a defining element of the cruise experience. The welcome and personal service in a restaurant, the care and empathy demonstrated by a room attendant, and the friendly willingness of an assistant purser to turn a potential problem into a resounding success are all momentous and memorable. The effect of great customer service can be to create loyal customers. However, the opposite is also true—bad customer service can turn customers away. In essence, customer service is a form of public relations (PR) directly involving interaction with the passenger (Kudrle and Sandler, 1995). Customer service involves high-level communication and an ability to tune in to the needs of customers.

Johns (1994) declares that businesses get the customers they deserve; in other words, the business that loses customers should look internally to identify the reason. All employees should practice customer service, although the front-line staff, in particular, should be trained to exhibit and implement the critical skills that can make a difference. Cruise companies build their brands on quality customer service to underpin the tangible elements of the cruise product. Invariably, customer service is itself built on certain critical components, such as ability to communicate, product and service knowledge, and interpersonal skills. The simplest single element that appears to play a disproportionate part in the customer service process is the ability of the individual staff member to smile at the right time (Clark, 1995). Clark believes this is because it is a human action that is universal, it sends a positive message, and it signals a willingness to interact in a friendly manner.

What Is Good Practice for Customer Service?

Customer service is subjective and is understood by the customer in a way that is particular to the individual. By its nature it can be highly personal and things can go wrong. An experienced front-line staff member will adapt what he or she knows to personalize each customer service event (usually a discussion between two parties) to create the best outcome that suits the customer and the company.

It is often difficult to remember that an unhappy customer is not making a personal verbal attack on the front-line staff member. Difficult customers who bring negative attitudes do not tend to create the best conditions for finding the positive outcome. Yet difficult customers are relatively common. People learn to complain by observation and practice. Our media-driven society creates many opportunities for people to hone these skills, and some revel in the opportunity to turn the skill into an art (Williams, 1996). Dealing badly with a negative customer can spread the negativity (to as many as five people, according to studies). Dealing positively with complaints on a cruise ship can turn a dissatisfied client into a satisfied one, generate positive PR, lead to increased sales, and help to improve working life for many personnel along the way.

Orientation for Customer Service

Creating the Right Attitude

Good front-line staff know when to talk and when to listen. Listening skills involve concentration and good eye contact (attending), careful use of nods and gestures as well as occasional probe questions or paraphrasing (following), and summarizing or confirming understanding (reflecting) (Clark, 1995). In addition, it is important to adopt the right attitude and to aim to be open minded before dealing with any contact. The setting is important, as it is better for the clients to be able to express themselves without concern and to contain situations that may create unwarranted audiences for potentially distressing situations. This can work both ways in protecting both the image and reputation of the customer and the company.

Letting the Customer Talk

It is always best to assume that the customer is truthful and to not try and find flaws or holes in the customer's story. It is important to remember that a complaint is a problem-solving issue and not a battle. The customer will believe she or he is right, and sometimes that position may appear unmovable. Talking allows the customer to communicate the problem to the person who should receive the information. It is important to enable the employee to gather facts, and it helps the customer to gain composure and settle down. This can defuse emotions and create a platform for negotiating the next stage.

Empathy

If the staff member expresses understanding for how the customer feels, he or she is not necessarily agreeing or admitting guilt in the face of a complaint. By using language and tone of voice with care, the staff member establishes the mood and can positively affirm the desire to deal with the situation by working with the customer. At this stage, the employee can check facts that are unclear and systematically make sense of what is said.

Problem Solving

When problem solving, it is bad practice to say "no." It is good practice to find out what the customer wants. Present options to the customer and examine and explore options to ensure that both parties can end up with an acceptable outcome. Options for dealing with the problem will depend on company policy. A justifiable complaint may require a gesture to be made that acknowledges the situation and provides some form of restitution. The amount of leeway will be a matter for policy to decide; for example, whether it is possible for a headwaiter to offer a bottle of wine to apologize for a serious delay in service or if a $100 credit can be provided by the senior assistant purser to a client's account to compensate for damage to personal property.

Most companies will not allow refunds to be offered on board and will expect that a customer file be sent to the appropriate department ashore so that any claim can be properly considered. Gestures of goodwill on board involving consumables or spending on board can be judged against the fact that while the selling price was not achieved, the loss in real terms is only the cost of the potential sale.

Follow Up

A complaint may seem traumatic for the staff but it must be remembered that it is possibly equally or more traumatic for the customer. In addition, the complaint may draw attention to a symptom of a larger problem. In this respect, the company should welcome all complaints and the front-line staff should be open in their gratitude to the customer. Any action point should be followed up without delay. Even if no follow-up contact is suggested, it is still good practice to get in touch with the passenger to inform them about what you did and to check that all is well.

Action and Resolution

This step involves recording the problem and creating a log of events to chart how the problem was resolved. It may be important to disseminate details of the problem more widely so that appropriate colleagues are informed and to monitor the situation more closely. Patterns or trends may exist that are not widely understood. Many problems are a matter of poor communication, and most suggest lessons for better practice. It is better to prevent problems than to aim to repair them (Crosby, 1996).

Important Exceptions

If a client departs from what is held as reasonable behavior and starts to threaten abuse either verbally or physically, policy should dictate the next step. This may require summoning a senior officer or immediately calling for help. No employee should feel that he or she is open to abuse. Employees should feel secure and safe while doing their jobs. Fortunately, such incidents are extremely rare.

When speaking with a customer, personnel should be careful not to use certain inflammatory phrases. Depending on the situation, these can include:

- You must be mistaken.
- I can't help you, or I don't know.
- Calm down, or Don't shout.
- That's never happened before.
- It wasn't me.
- Sorry, that is not my problem.

The situation may also become inflamed if members of the customer service staff look bored, are distracted by a colleague when in the middle of a complaint, or adopt a patronizing tone of voice. Staff members should aim to provide realistic promises and identify worst-case scenarios in a timely manner. Customers are impressed by service improvements that address a stated target. It is good practice to underpromise and overdeliver.

Summary and Conclusion

This chapter has investigated a number of theories related to service quality and has applied these theories to the context of cruise ships. Good managers make the difference in service quality, but the complex environment of a megacruise ship is such that there are numerous fault lines that can emerge. Growth for a cruise company can be positive, exciting, and rewarding; yet, critical qualities can be threatened when new staff members are poorly trained and prepared, experienced staffers are stretched or asked to do more, and corporate planners ignore the operational tensions arising from growth.

Glossary

Audit: To undertake independent review and examination of system records and activities in order to test their adequacy and effectiveness.
Baby boomer: A person who was born in the generation between 1946 and 1964 when there was a growth in the birth rate.
Empathy: Being able to put oneself in another person's position so as to understand what it feels like.
Compensation: Something (such as money) given or received as payment or reparation.
Induction: Training intended to introduce new staff to the workplace.
Lines: Queues of people formed to facilitate a human process fairly.
Orientation (at the beginning of a cruise): Finding out where everything is so the passenger does not get lost.
Persona (marketing term): Psychographic term for a type of customer.
Specification: The design and template that establishes minimum and maximum standards and parameters for a product or service.
Tangible: Perceptible by the senses, especially the sense of touch.

Chapter Review Questions

1. What does *quality* mean?
2. What is the difference between a service and a product?
3. What is TQM?
4. What are the options for managing tipping?
5. Describe the main stages of dealing with a customer.

Additional Reading and Sources of Further Information

http://www.mazur.net/tqm/tqmterms.htm: Dictionary of TQM terms
International Journal of Service Industry Management
Journal of Retailing and Consumer Services
Managing Service Quality Journal
The Service Industries Journal
The TQM Magazine

References

Clark, M. (1995), *Interpersonal skills for hospitality management*. London: Chapman and Hall.
Crosby, P. (1996), *Quality is still free making: Quality certain in uncertain times*. New York: McGraw-Hill.
Dale, B. G. (1999), *Managing quality*, (2nd ed.). Hemel Hempstead: Prentice Hall.
Deming, W. E. (2000), *Out of the crisis*. Cambridge, Mass.; London: MIT Press.
Douglas, N., and Douglas, N. (2004), *The cruise experience: Global and regional issues in cruising*. Frenchs Forest, Australia: Pearson Education.
Harrington, D., and Lenehan, T. (1998), *Managing quality in tourism: Theory and practice*. Dublin: Oak Tree Press.
Harris, N. D. (1989), *Service operations management*. London: Cassell.
Johns, T. (1994), *Perfect customer care*. London: Arrow Business Books.
Juran, J. M. (1980), *Quality planning and analysis: From product development through use* (F. M. Gryna, Trans. 2nd ed.). New York: McGraw-Hill.
Kudrle, A. E., and Sandler, M. (1995), *Public relations for hospitality managers*. New York: John Wiley and Sons.
Lovelock, C. H. (1992), *Managing services*. New Jersey: Prentice Hall.
Office of Quality Management (2005), Quality Bytes: managing perception points to make service perceptions last, 4 Aug 2005, from http://www.nus.edu.sg/oqm/news/qbytes/archive/issue0013/
Peters, T. (1987), *Thriving on chaos*. London: Pan.
Tse, E. C. (1996), Towards a strategic total quality framework for hospitality firms (M. D. Olsen, Trans.). In M. D. Olsen, R. Tear, and E. Gummesson (Eds.), *Service quality in hospitality organisations* (pp. 316). London: Cassell.
Williams, A. (2002), *Understanding the hospitality consumer*. Oxford: Butterworth Heinemann.
Williams, C., and Buswell, J. (2003), *Service quality in leisure and tourism*. Wallingford: CABI Publishing.
Williams, T. (1996), *Dealing with customer complaints*. Aldershot: Gower.
Wright, J. N. (2001), *The management of service operations*. London: Continuum.
Wyckov, D. D. (1982), New tools for achieving service quality. In C. H. Lovelock (Ed.), *Managing services: Marketing, operations and human resources* (2nd ed.). Hemel Hempstead: Prentice Hall.

8

Managing Food and Drink Operations

Learning Objectives

By the end of the chapter the reader should be able to:

- Identify how food and drink operations are managed on board a cruise ship
- Consider issues relating to stores and supplies
- Understand the systems approach to managing food and drink
- Reflect on organizational issues on board

There are few more complex or demanding food and drink operations than those that are operated on a high-specification contemporary cruise ship. Yet, at a time when comparable shore-based operations are locked in a struggle to achieve the highest standards of output within an environment that demands cost cutting, deskilling, and centralized production, the caliber of provision of food and drink on these types of cruise ships is a veritable beacon of excellence.

Kirk and Laffin (2000) describe the unique problems faced by those who are involved in "travel catering" and, more specifically, in cruise ship catering. Below are some of the points they cover.

- The quality of the food on board is critical to the success of the cruise, yet the food product is not the main reason to choose this kind of vacation.
- The price of the vacation includes the provision of food and (sometimes) drink.
- Cruise ships operate a variety of restaurants to meet diverse customer requirements.
- There may be complicated logistics involved in setting up a supply chain.
- The facilities for dining are managed to achieve the highest possible standards.
- Passenger-to-crew ratios can be low, so the levels of service are potentially high.
- The crew is expected to work long hours seven days a week for several months at a time.
- Crew rewards can be high because of tips, tax-free wages, and the opportunity to travel.

Kirk and Laffin (2000) note that the layout of galleys aboard tends to be similar to more conventional kitchens ashore. They further identify that space is less of a problem than it is for other types of travel, such as air travel. However, new generations of cruise ships are forging ahead with galley design to the extent that comparison with shore-based operators is fast becoming anomalous. Galleys on cruise ships are constructed to effectively meet the unique demands of producing high-volume, high-caliber meals safely, effectively, and consistently.

In some respects, the constraints of the food and beverage operation on a cruise ship at sea provide certain compelling advantages. The ship's staffing levels (the ship's complement) must be geared to meet statutory legal requirements and to meet operating needs at full capacity. Staffing levels cannot be suddenly reduced or increased. The optimum levels of staffing need to fit all circumstances. Planning needs to take into account any contingency issues that could be affected by the ship's itinerary. There are serious implications if a ship is not appropriately serviced and in receipt of consumables. The logistical exercise for supplying stores to a ship is thus influenced by a need to ensure that operations are not jeopardized by an unexpected occurrence such as a technical delay, change of itinerary, or bad weather.

Supplies and Services

The world's famous transatlantic liners were designed to be self-contained entities. The ships were away from home ports and at sea for lengthy periods of time so, in effect, the ships were floating cargo as well as passenger carriers. Frequently, ships called in to certain ports to purchase indigenous goods that were then stored and taken back to the home port for distribution to other ships in the fleet. An example of this practice would be a ship calling into Auckland for New Zealand lamb. Storage on these vessels was designed to meet these needs with cavernous freezer storage, refrigerated spaces, and dry goods areas located on lower decks close to loading bays.

Contemporary cruise ships are different. They are designed with certain features in mind. Usually, they are not intended to sail in difficult sea conditions, and the maximum space is provided for passenger areas to create as much income from cabins or staterooms, bars, and other revenue-generating areas as possible. The primary purpose of the ship is to generate profits for the company. Therefore, in designing the ship, care is taken to ensure that space is allocated accordingly. If the ship is designed for 10-day cruises, storage is designed accordingly. At the beginning of such a cruise, the ship's storage areas are stacked to the limit and, at the end, the stock holding will be minimal.

The business of managing a ship's food and beverage supplies can be highly technical and specialized. For a major cruise company the process begins at the head office and involves consultation with a number of key professionals aboard and ashore. Planning involves reflecting on prior patterns of consumption, identifying changes to expected routines, planning menus for different passenger types and itineraries, and forecasting needed quantities. Contracts are offered on the basis of ability to supply, quality, and price. Because of the size of the contracts, this business is highly lucrative and attractive, but the scale of operations invariably means that some suppliers are unable to meet the criteria for the contract tender.

A head storekeeper or stores manager is responsible for stores on board and is often aided by an assistant and an administrator. In addition, ships frequently employ a cellar master who is responsible for beverages. These employees report to the food and beverage manager or equivalent, and work closely with the head chef or the bar manager. The routine of receiving and storing goods generally begins on arrival at the port of departure. This may be the home port or the port that is selected for embarking and disembarking passengers for a series of cruises.

On the dockside, goods are held in a container that is sealed by the supplier and checked by customs officers prior to the arrival of the ship. When the ship ties up and has been cleared by customs or port officials, goods can be loaded. This is usually done with forklift trucks and pallets, although conveyor belt systems are often used on board. Most large vessels are loaded via doorways that are located at the quayside level. These are frequently referred to as "gun port doors," an historical nautical term. On the quayside, stevedores supervise a team of dockside laborers to deliver goods to the ship, and on-board general assistants are deployed to ensure that the various supplies are stored correctly. Stores managers check the items for accuracy and quality. Goods can be rejected if the quality is below specification.

Most cruise companies operate a computerized stock management system that allows goods to be controlled accurately. Requisitions can be undertaken electronically from the chef's office or from the bar manager's office and stock is then collected and checked against the requisition. The stock management system allows the inventory in the stores, cellar, and bars to be easily checked against the record to ensure there are no discrepancies.

Stores managers are responsible for the safe and accurate management of stores in their areas. Stores assistants, and anyone involved in handling goods, must be trained to undertake this task safely. The volume of stores arriving can create pressure to ensure that the storing process is carried out as quickly as possible, that the required health and safety considerations are met for personnel lifting heavy goods without hurting themselves, and that items are handled in an appropriate, safe, and hygienic fashion. Goods must be stacked or stored securely so that they do not get damaged or cause accidents if the ship moves abruptly or if the general conditions of storage are otherwise inadequate. Goods that are perishable must be rotated to ensure that waste is minimized and that the highest quality of produce is supplied.

Stores such as caviar and vintage wines require careful treatment, because of their prestige value, and may be stored more securely. Other items, like fresh vegetables, may be stored in order to maintain or develop their readiness for use. The purchasing specification provided to the supplier will outline the expected condition of the produce, such as the state of ripeness for fresh produce. When items are requisitioned from stores they are then transported to the next stage in the process. Drink products are distributed to bars or the restaurant dispense bar. Food items go to preparation rooms in the galley, the bell box, or the main galley.

Food Production and Service Delivery Systems

According to Ball, Jones, Kirk, and Lockwood (2003a), hospitality organizations provide excellent examples of the workings of systems theory, a theory that helps to explain complex situations (Kirk, 1995). Systems can be subdivided into "hard" systems, which are technologically based, and "soft" systems, which have to do with people. Ball et al. (2003a) describe a system as "a set of components and the relations between them, usually configured to produce a desired set of outputs, operating in the context of its environment." The study of systems enables managers to deconstruct a process and potentially make improvements or introduce a new system, with due regard to planning, structure, and relevance.

The authors identify a number of principles that they feel are valuable when reflecting on systems theory (Ball et al., 2003a).

- Systems are most at risk during periods of change because there appears to be a feeling that change is generally resisted.
- Systems and interacting systems are complex, and some elements of a system appear to be more risk prone, or "dispersive," while others are stable or "cohesive." It is important to develop a balance to ensure that the dispersive elements do not disrupt and, thus, ultimately undermine the system.
- Systems must adapt at the same rate as the environment changes in order to maintain cohesion and balance. The authors suggest that, in complex settings where systems interact with other systems, a higher degree of stability occurs when the interaction is between a larger number of and more varied types of systems.
- Systems have a limitation, in terms of variety, that is predicated by the environment.
- Systems may not be totally standardized across a group, yet they may be stable and unique within individual settings.
- Interconnected systems appear to undergo a cyclic progression from the point where the system variety is generated to the emergence of a dominant approach, the suppression of variety, the breakdown of the dominant approach, and the re-emergence of surviving variety.

Ultimately, systems theory develops a logical approach that, for food and beverage operational management on the cruise ship, may be described as the food production system (involving food preparation, production, holding, and transportation) and the food and drink service system (involving food service and dining, clearing and dishwashing, and bars). According to Davis, Lockwood, and Stone (Davis et al., 1999), food production is the process concerned with converting raw, semi-prepared, or pre-prepared materials into ready-to-consume items. The system's effectiveness and efficiency are reflected in both the relationship between inputs and outputs—waste, energy efficiency, and labor efficiency—and service delivery factors—customer satisfaction, perception of quality, and service issues (Ball et al., 2003a).

The production system on a cruise ship has three key elements: the policy relating to catering on board (variety of outlets, variables in terms of demand, type of catering operations, timing of services), the menus (style and quality factors; implication for production if à la carte or table d'hôte; number, variety and standard of dishes; preparation of food noting standardized recipes, volumes, implication for service, and portion control) and galley or kitchen design.

Contemporary cruise ship galleys appear, at first sight, to be vast areas of stainless steel populated by chefs working in gleaming white uniforms to produce culinary works of art as if by some kind of magic. The production system on most ships is a derivation and development of the traditional French *partie* system, where *mise en place* or preparation is done in relative isolation, only to come together at the time of service in an orchestrated manner. The *partie* on cruise ships is less reliant on French culinary terminology, such as *garde manger* for the larder or *saucier* for the soups and sauces, but the compartmentalization is nevertheless effective, using terms more relevant to the type of menu (such as fish, pastry, butcher, soups, and pasta). This approach is less visible on cruise ships because the preparation stages take place in closed areas to prevent contamination and to maintain high levels of hygiene control.

The main galley is designed to enable the assembling of the prepared items in a logical manner so that they can be collected by waiters and served efficiently to customers without any degradation of quality in terms of appearance, temperature, or taste. The location of the galley is important to facilitate this objective. It is best located adjacent to the preparation rooms and the restaurants that it is servicing. Ideally, it should be on the same level, although lifts are frequently used for transporting food from preparation rooms to satellite restaurant galleys on ships where there are bistro or specialty restaurants. In these cases, the lifts are prioritized for the specific use of carrying foodstuffs and treated accordingly. The location aspects also apply to crew dining and the location of washing and storage areas.

Figure 8.1: A table setting on the '*Queen Elizabeth 2*'

Galleys require good and effective ventilation to ensure that working conditions are comfortable, condensation is minimized, and cooking odors are controlled (Ball et al., 2003a). In addition, the galley requires cold water supplies for drinking, cooking, and washing up and hot water for other various purposes. The galley must have adequate drainage; a safe, user-friendly, durable, and hygienic floor surface; suitable lighting; and hygienic, safe, and easy-to-use work surfaces.

The food and drink service system is concerned with delivering food and drink to the customer (Davis et al., 1999). The way it is done is vital, of course, in satisfying the customer and demonstrating the inseparability of food service from food production and vice versa. Indeed, on some cruise ships there is a blurring of the two processes, as can be seen in the Grill Room on the *'QM2'* where flambé (cooking and flaming of food), is done in theatrical fashion at the table. This is sometimes referred to as *guéridon* work, after the name given to the trolley and lamp. Safety devices on the ship prevent the use of gas fuels for cooking at the table. High-powered electric elements are used instead.

Food service systems take account into several factors: *timing* (when the customer wants the service or when the service is scheduled to take place); *location* (restaurants, buffet service, room service); *customer needs* (silver service, semi-silver service, plated service and *degree of social interaction* between staff and customer). Service system types are usually hybrid versions that have emerged from historical service methods. Thus, in formal dining areas, plated service or semi-silver service may be by waiters and assistant waiters used to serve food to the table. In informal buffets, a combination of table service and self-service may be used. A hybrid style is a service style that has been adapted to suit customers and to create a service routine.

The service process involves *mise en place* (preparation), when the room is prepared and thoroughly cleaned, tables are laid, the cutlery and glassware polished, the *serviette* (napkin) folded, the cruet set filled, and the appearance of the table setting carefully checked to make sure it is uniform, attractive, and appropriately prepared (see Figures 8.1 and 8.2). Waiters and assistant waiters work from sideboards that contain all necessary items during the service, including water, glassware, spare cutlery, bread rolls, various sauces, and a pepper mill. These items also need preparation before service

Figure 8.2: The dining room on the *'Queen Elizabeth 2'*

begins. Wine waiters (*sommeliers*), prepare any items they may require, such as decanters for wines that throw a sediment (which is not common), spare glasses, liqueur trolleys, and ice buckets. The buffet also needs preparation to ensure that customer tables, buffet areas, beverage points, and clearing areas are ready for service.

Service follows a pattern dictated by the arrival of guests (whether one or two sittings or flexible dining). Experience informs the managers as to the type of traffic flow to expect within the time constraints of the meal. Often, customers eat early if the ship has been in a port with a late afternoon departure or if a show is scheduled that is attracting interest. The meals on board most cruise ships tend to be included in the price, and it can be difficult for staff to treat requests for a customer's third serving of lobster without batting an eye. Some restaurants on cruise ships have a certain cachet or prestige value because, as on the *'QM2'*, it relates to the grade of cabin and the menu and service is available on a similarly graduated scale. Other restaurants are made available at a supplement to reflect a specialty. The menu may be designed and branded by a celebrity chef or the style of restaurant may be unique in some way.

The bar service system is designed for the "purpose of dispensing and consuming alcoholic and nonalcoholic beverage" (Ball et al., 2003a). Most cruise ships with an American clientele adopt an approach to control the bar area that is in keeping with licensing in the United States. Thus, anyone under 21 is unable to purchase and consume alcoholic beverages. The same rule is not applied on ships with a predominantly Italian, Spanish, Australian, German, or British clientele, where anyone who is 18 or over can be served. Alcohol is potentially dangerous, and the bar staff is trained to comply with company policy regarding the sale of alcohol to safeguard passenger interests.

The sale of drinks involves a bar with a counter. The back area of the bar is used to display the various beverages that are available, while the under-counter area houses items and equipment such as glassware, preparation areas, refrigeration units, ice containers, cocktail making equipment, and sinks. The theory is that functional items should be hidden from view while items for sale should be easily seen. Different types of bars are fitted to meet different purposes. A sports bar may have greater volumes of draft beer sales compared to a champagne bar, and a cocktail bar may serve more after-dinner cocktails than a pool bar. Seating and layouts are also highly individual. It is normal practice to design bars that have a view of the front or back of the ship (commonly named the "lookout" bar, the "crow's nest" bar, or the "ocean" bar), and the design can frequently use split levels to maximize the number of seating and table configurations that are in "pole position". Some bars are designed with intimate areas, others with "see and be seen" areas. Cruise ships tend to favor table service (rather than service at the bar counter) as a way of maximizing sales, making sure the customer enjoys a feeling of being looked after and keeping crowds or lines to a minimum.

In common with both the galley and the restaurants, the bar service system also involves *mise en place*, with the preparation of beverages, glassware, displays, ingredients, decorations, and garnishes. The general preparation of the bar area includes cleaning and polishing tables and counter surfaces, and positioning the appropriate items on the tables (various bar lists, promotional tent-cards, and drink coasters), the bar (display areas), and the bar counter. The bar is a focal point and is highly effective in generating sales. Sales tactics can include careful positioning of premium products, promotional displays, a raised floor area behind the bar to highlight the bar staff and give them an enhanced view of their trading area, and raised lighting behind the bar to increase the impact and visibility of the area.

Organizing People, Products, Processes, Premises, and Plant

It is a function of food and beverage management (Davis et al., 1999) to coordinate the organization of teams, to monitor and review the production and service of products, to implement and evaluate

Figure 8.3: Food preparation in the galley

processes as laid down by standard operational procedures, and to maintain an overview of operations that affect premises and plant so as to achieve effectiveness and efficiency.

Food and beverage personnel operate in teams (see Figure 8.3). The required activities within the food and beverage areas are such that greater job satisfaction and productivity can be achieved through teamwork (Ball et al., 2003a): less repetition (when tasks are shared, there is the potential for a higher degree of interaction and communication in the work area, and individuals in teams can support each other), more efficiency (production systems and service routines can deal with higher units of output when people work together, health and safety can be managed and monitored more effectively within an open team environment), and greater effectiveness (quality control is overt in a team setting, and the aforementioned support can lead to higher levels of competence and shared good practice). Correspondingly, nurturing the teams ensures that working relations are maximized and company goals are achieved.

In monitoring and reviewing the production and service of products, the food and beverage manager acts as the arbiter of quality; providing a lead in raising standards; creating clarity of purpose and focus for fellow managers, supervisors and operatives; and maintaining a highly visible presence as a key individual within a team. Complex food and beverage operations that have multiple outlets and continuous, frequently high-volume production and service routines require constant attention to ensure that high standards are maintained. The part managers play in this sense is pivotal to assuring quality. The manager is a *conductor*—loyal, caring, and sensitive to the component parts of the operation; a highly proficient *communicator*—capable of motivating teams and individuals, yet unyielding in an expectation of high quality and health and safety compliance.

Systems evolve within a setting and, in many respects, the evolution can be positive. However, it is the food and beverage manager's role to assess processes in consultation with key colleagues so as to address potential problems or issues. Because food and drink systems are frequently integrated and reliant on the work of groups of people, food and beverage managers must be diligent in observing operational routines to identify emerging or inherent flaws or risks in these routines. A manager in this position can only make decisions about remedying problems from a position of strength, experience, and knowledge. Standard operational procedures can provide a framework that guides the food and beverage manager and helps to establish minimum standards, but interpreting it is a human activity that requires conscientiousness.

The organization of the food and beverage department focuses on a unique, carefully designed setting that is, in the case of the service areas, a place of work and entertainment and, in the case of a production area, a potentially dangerous and high octane environment. The food and drink service emulates theater. The staff members are actors who integrate with the audience, performing their duties and, if successful, creating customer satisfaction. Managing the environment is complicated by this need to maintain the theatricality of the setting. The mechanics of production and much of the work associated with service are hidden from view in a desire to present an experience that is desirable and appreciated by the customer. The manager is responsible for making sure the environment is not compromised by rogue practice or by deterioration of the premises or plant. The efficient operation of equipment is paramount to preventing damage and unnecessary cost. This aspect of a manager's job is termed asset protection (Verginis and Wood, 1999).

Customer Demands and Operational Capabilities

Ask food and beverage managers how they know what the customer wants and, invariably, the reply will be along the lines of "years of cumulative experience." Yet dining experiences are changing with the new demographics now on board and with contemporary trends in eating out. With cruise brands targeting new market sectors, the product designers are being faced with different challenges to satisfy customer demands and expectations.

Roy Wood (2000a) doubts that the hospitality industry knows that much about people's eating needs today. Yet the cruise industry seems to get it right. The individual brands develop products that are elements of the brand identity. Some British vessels may emulate a more traditional dining pattern, with the type of food seen in high-class, sometimes more formal restaurants (as on the Cunard Line). Some US-based vessels adopt a casual pseudo-Italian approach to dining, reflecting a friendly and informal approach to meal time (as on Princess Cruises). The origins of the "product" appear to be traceable. With Cunard, the "Britishness" of the dining experience, and indeed of the cruise, is the unique selling proposition (USP) that attracted UK and US passengers to the early cruise ships. The tradition, once established, is protected (often fiercely) in order to establish and maintain the company's branding or perception of identity.

This same kind of analysis can be done with Princess Cruises. The company absorbed "Sitmar," an Italian brand, during what was an important formative period for the cruise company. The resultant influx of Italian staff, and their establishment within the new "Princess" brand, led to the development of the Italian-American emphasis on food and menus. Subsequent growth has come from within the company, with the promotion of highly skilled and experienced food and beverage managers and the consolidation of the strategic vision for delivery of food and beverage as part of a brand identity.

The product and product design have, for many brands, emerged as an evolving developmental process. New ideas can be tested on ships before being rolled out across the brand. Cultural elements,

Figure 8.4: The restaurant on the '*Aurora*'

including ports on an itinerary, predominant crew nationality, and predominant existing and new passenger nationalities, all play a part in creating the changing setting and in stimulating new ideas for developing the food and drink on offer. Consideration of the passenger service questionnaire and the high degree of interaction between passenger and staff means that awareness about perceptions of the product is indeed a matter of cumulative experience.

Operational capability is tied into the basic concepts associated with the notion of food and drink as elements of the brand. Style of service, skills acquisition, knowledge and learning, and the need to deliver food and drink to standard and within budget form the other elements of the equation. The number of people who serve in a given time is modified by the service routine and the product specification.

Several types of operational problems can occur when food and drink planning is changed. For example, establishing an "al fresco" dining area seems to create an excellent passenger option that capitalizes on the location of open areas on decks and provides a dining alternative. Typically, however, the distance from the finishing galley to the restaurant needs to be surveyed to make sure the food can be delivered safely and to standard.

Control Actions for Food and Drink Operations

According to Davis et al. (1999), there is a need for systematic quality management in the provision of food and beverage which should include inspection, analysis, the design of systems to prevent problems, and a team-focused quality-oriented approach. Quality control can stem from the way that a team embraces the critical elements that help the company to produce and serve food and beverages to meet or exceed customer expectations safely and within budget. For a bartender, quality may be reflected in the taste and appearance of the product coupled with the ambience of the room and the service skills involved in delivering the product to the customer. The quality may well be undercut if portions are inaccurate and profitability undermined.

The balance on board a cruise ship is to produce food without waste. This balancing act is performed by producing food to a pattern based on historical data and supplying products to customers to meet reasonable expectations. Portions on the plate are designed to look good and to satisfy expectations. Some customers may eat more than others, but that is managed by delivering acceptable portions that can, if necessary, be increased on request.

Food Safety, Health and Safety, and Consumer Protection

Production and service of food and drink can be compromised by poor safety and hygiene. Therefore, cruise ships are vigilant in promoting best practices for the management of food and beverage areas. In addition, port health authorities undertake inspections that are highly visible and can be important for a cruise ship's reputation. In the United States, the Centers for Disease Control and Prevention have operated the Vessel Sanitation Program (VSP) for almost 35 years. This government agency biannually inspects water, food, spas, pools, employee hygiene, and general cleanliness on board cruise ships that carry 13 or more passengers, scoring them on a 100-point scale. In addition, the CDC provides an annual summary known as the "Green Sheet" to highlight issues and make recommendations.

The CDC examines potable (drinkable) water supplies to make sure the storage and transfer equipment is clean and that the water is microbiologically analyzed to ensure it is safe for consumption. Pools and spas are checked to ensure they are safe and well maintained. Personnel are checked to highlight infections, management of hygiene, staff knowledge relating to hygiene, and monitoring of food safety. In addition, compliance with recommendations for safe hygiene practices and training plans is examined.

Food products are inspected, storage or holding temperatures examined, thawing practices noted, cross contamination checked, and general practices such as the protection and storage of food, labeling, and dispensing monitored. Equipment such as food contact surfaces, production equipment, washing equipment, and utensils are all examined. Uniforms of chefs and food handlers, and cloths or towels are inspected and hand-washing facilities carefully examined. The inspection goes into considerable depth, even checking bulkheads (ceilings) and deck heads (floors). Medical records are also noted as part of the process. Any ship scoring less than 86 is deemed to have failed.

Planning Wine Lists

Wine is perceived as a highly desirable adjunct to, or component of, the meal experience. The consumption of wine appears to be more socially acceptable than, say, drinking beer or spirits because the product carries connotations of knowledge and connoisseurship (Ravenscroft and van Westering, 2001). Therefore, there is much to be gained in selling wine, either as a beverage during social interaction or as part of the meal experience.

That said, for many, the subject creates fear. Prial (1990) recognizes that some people can be intimidated by the prospect of buying wine and, as a result, they avoid the process. Wine as a subject is laden with theory and, for those in the business of selling wine, there are critical aspects to note when designing wine lists and making sales. Cruise brands have the potential to sell large quantities of wine, and the very diversity of wine as a product creates potential for matching products to settings and to client types. The following section is intended to provide some guidance relating to the selection of wines for selling on board.

Cruise brands are in a strong position because of their employees' cumulative knowledge about which wines to stock and sell. Yet, like food, a broad range of elements can affect both the trendiness and acceptability of wine types. These elements can include passenger diversity and the variability by cruise of aspects such as age and demographics, socioeconomic background, wine knowledge, and social and cultural circumstances.

A starting point could well consider the logic and practicality of constructing individual wine lists for each cruise. Doing so might encourage greater sales by ensuring that products are made available and sold in line with predicted demand. The alternative approach is to offer as wide a range of wines as possible to suit the market in its broadest sense. Constraints on range include the availability of storage space, the potential for being left with redundant or deteriorating stock, the availability and continuity of supply, the complexity of managing a wine cellar on board a ship, and the investment cost of holding large supplies.

The logical progression of this argument suggests that a contemporary cruise brand with a definable clientele may opt to develop a standard wine list that can be amended if required. The wine list forms the basis of supplementary lists (for speciality restaurants) or offers (wines of the day) and is the product of close consultation between the cruise brand purchasing team and an identified wine supplier. Certain generalities hold true when wine is purchased in large volume. Continuity of supply is predicated upon quantity of production. The wines on a wine list that is replicated across a fleet will have been sourced from wine producers who are in the business of producing high volumes. This implies that smaller producers, who may well have excellent reputations for the quality of their product, generally cannot be considered.

Selecting wines is often a matter of dividing them by color (red, white, or rosé), by production method (still, sparkling, or fortified), by body (full to light), and by acid and sugar content (dry to sweet). This process helps to create a range that is sufficiently diverse to meet clients' needs and expectations. Some cruise brands may be required to stock products that have a higher prestige value, which might translate into the need to hold vintage champagnes, expensive burgundies from France, cult wines from California, or famous wines from Australia. Inevitably, however, the sales patterns for most cruise brands will reflect a balanced stock-holding that is biased toward less expensive products.

Points to note when planning a wine list follows:

- The list should have a logic that is easily understood by the client. Categories of whites and reds may be sufficient to delineate initial options, followed by country or area. More traditional approaches can be taken to list countries by significance from a "sales on board" perspective. Shorter lists are easier to digest, but the list may lack impact or image if it is too constrained.
- Wines should be selected that are consistent in quality and availability. Quality is implied by the information on the label that, depending on the wine and country, could identify and guarantee the derivation of the contents. Quality is subjective, so wine-tasting samples can help customers make critical decisions. When purchasing from volume suppliers, it is useful to assess quality regularly to ensure that standards are consistent.
- Vintage (the year of the grape harvest) is important for most wines. The year can indicate the potential for longevity or may be a warning about a potential "sell-by" date. Some wines, mainly reds and some sparkling wines such as vintage champagnes, are capable of aging because they are made in such a way that they will mature in the bottle. Other wines, mainly whites and some light bodied reds, are best consumed within one or two years of bottling.
- Both wine connoisseurs and novice drinkers like to recognize familiar wines. These wines can create a point of reference and help to make the novice more comfortable and the connoisseur more trusting. Wine lists can be a mix of these familiar names together with others that are deemed reliable. Wines can be named after grape varieties such as riesling, sauvignon blanc, chardonnay, pinot gris, viognier, pinot noir, shiraz (syrah), or cabernet sauvignon (to name but a few). This is a common approach that can help consumers who feel able to identify wines by their varietal nature. Others are named by a brand or place, such as Beaujolais (a town in Burgundy, France), Sancerre (a town in Loire, France), Cuvée Napa (a brand incorporating Napa Valley, California), Barolo (a town in Piedmont in Italy), and Villa Maria (a brand from New Zealand).
- The selling price (SP) is important. Some clients may not be inhibited by price, yet in most cases clients will have an understanding about the relationship between value and SP. Compared to food, it appears that little is done to wine in terms of service to enhance the value. For some, this can mean that they have a perception that the product can represent poor value if much above retail pricing. Some cruise passengers are aware that previous cruising price structures offered drinks on board to achieve lower margins because the basic cruise price once returned considerably higher margins than is currently the

case. Wine products can create good returns, but investment in glassware, equipment, and staff need to be noted.

- The wine list should achieve a balance to ensure that the wines offered represent an amalgam of products by type, origin (noting connections with the itinerary), and price range, noting client types, and should offer alternatives such as half bottles (for those who may not wish to consume a whole bottle), wines by the glass, and low or no alcohol wines. It is also common practice for passengers to order a bottle of wine but then send it back for storage rather than purchase wine by the glass or half bottle.

Finally, the skills and knowledge of those who work with wines are worthy of discussion. A good wine waiter or sommelier has accrued years of knowledge and experience. The subject is weighty, but interest in the topic of wine among the general public remains high. Wine knowledge and skills require investment in order to meet possible demand on board and to ensure effective wine sales. Some products require a higher degree of care than others. Interestingly, these products tend to be those that require one thing a cruise ship may struggle to offer—a stable cellar. It is still possible to stock products that require processes such as decanting (to remove sediment from aging wines), although it is more common to avoid this by stocking wines that do not require such treatment.

Case Study: The Executive Corporate Chef and Menu Planning

Planning menus and food on the '*Star Princess*' is a systematic exercise. For the cruise that this case study relates to, there is a 12-day cyclical menu. The company has menus that can span cruises that are 3 to 30 days in duration. Menus are sensitive to the nationality of passengers so as to satisfy their preferences. The main restaurants on the '*Star Princess*' serve exactly the same food and therefore, for a 12-day cruise, there are menus for twelve breakfasts, twelve lunches, and twelve dinners. The actual menus are laid out in exactly the same way so that the passengers are not confused at first and so that they can quickly become familiar with the format.

On the lunch menu, the first page presents the beverage suggestions that can include cocktails, wine by the glass, beer, and mineral water. This is followed by appetizers and a soup, and then followed by a salad and "always available" items such as hamburgers and cheeseburgers. The second page includes "favorites" in the form of two pasta dishes, followed by a range of main courses that include one fish item, one meat item, one poultry item, one main course salad, and one vegetarian dish. Desserts follow with three ice creams, two frozen yogurts, gelatin, and cheeses.

The dinner menu has a different format. Page one presents a complete healthy three-course meal. This menu indicates that the proposed items are lower in fat. This is followed by a complete six-course vegetarian meal. Standard dishes that always appear on the menu are listed, such as classic Caesar salad and plain grilled fish, chicken, and beef. Page two begins with the appetizers, or hors d'oeuvres, which include three food items that may be fish, meat, vegetables, or fruit. The soup and salad section offers three soups (one cream, one consommé, and one chilled soup) and a salad dish. There is then a main course pasta dish and a range of main courses including fish, shellfish, an alternative meat (such as veal, pork, or lamb) a red meat dish, and a poultry dish. Desserts are presented on a separate menu that includes a selection of ice creams, pastry items, fruit items, hot dessert, and cheese.

The crew also must be fed. The executive chef plans a 30-day menu that caters to their needs. The menu is constructed with input from nutritionists and crew representatives, which ensures that crew members who may have a special preference or diet because of their nationality, cultural background, or religion are appropriately catered to. The crew menu includes a standard hamburger, a vegetarian dish, cold cuts, soup, rice, pasta, and two main course dishes (meat or poultry) in the evening. At lunchtime fruit, cheese, and ice cream are offered and for dinner a dessert dish is offered in place of ice cream.

Case Study Questions

1. What is the reason for a cyclical menu?
2. What critical issues inform
   a. passenger menu design?
   b. crew menu design?

Summary and Conclusion

Food and drink are vital elements for a cruise. Passengers put considerable emphasis on the provision and perceived quality (and sometimes quantity) of food and drink, both in terms of feedback provided in questionnaires and when relating accounts to friends, acquaintances, or family. This is an area of provision that is not outsourced (as can be the case for operating tours ashore). It requires very high staffing levels (the majority of crew work in this department) and it deals in high volumes (the largest ships are feeding in excess of 4,000 people at least three times a day). This type of operation must be systematic, as carefully planned as a military campaign, and consistently monitored to ensure that it is under control.

Managing food and drink is highly specialized, requiring a combination of skills and knowledge. Such individuals should have an appreciation of the art as well as the science of producing food and drink. The food and beverage manager is like a theater director, ensuring that people, plans, and conditions are in place for a finely tuned performance that meets or exceeds the audience's expectation. Finally, the manager has a responsibility to assure that the provision of food and drink is achieved safely and within the budget. The next chapter considers a parallel role: that of facility management.

Glossary

Al fresco: Relates to dining on decks in the open air.
Bell box: A separate galley for room service on a cruise ship.
Caviar: The salted roe (eggs) of the sturgeon.
Champagne: Sparkling wine from the champagne region in France.
Galley: The kitchen on board a ship.
Stevedore: A port worker who loads cargo in the hold of a ship.

Chapter Review Questions

1. How and why do shipboard processes for producing meals differ from shore-side processes?
2. What comprises a typical food and drink production system for a contemporary cruise ship?
3. What is the US VSP and how does it affect food and drink operations?
4. What are the critical issues to consider when planning a wine list?

Sources of Additional Information

http://www.caterer.com/home/ *Caterer and Hotelkeeper* magazine
http://www.cruisegourmet.com/ Cruise cuisine

Decanter magazine
http://www.hcima.org.uk—HCIMA—Hotel Catering and International Management Association

http://www.hospitalitynet_org/index.html—Hospitality Net
http://www.chefmagazine_com/main_splash_chefww.asp—*Chef* magazine
Journal of Restaurant and Foodservice Marketing
http://www.porthole.com/—*Porthole Cruise Magazine*

References

Ball, S., Jones, P., Kirk, D., and Lockwood, A. (2003), *Hospitality operations—A systems approach.* London: Continuum.

Davis, B., Lockwood, A., and Stone, S. (1999), *Food and beverage management* (3rd ed.). Oxford: Butterworth Heinemann.

Kirk, D. (1995), Hard and soft systems: A common paradigm for operations management, *International Journal of Contemporary Hospitality Management*, 7(5), 13–16.

Kirk, D., and Laffin, D. (2000), Travel Catering. In P. Jones (Ed.), *Introduction to hospitality operations.* London: Continuum.

Prial, F. J. (1990), What Is It With Wine? *Journal of Gastronomy*, 6(3), 3.

Ravenscroft, N., and van Westering, J. (2001), Wine tourism, culture and the everyday: A theoretical note, *Tourism & Hospitality Research*, 3(2), 149–163.

Verginis, C. S., and Wood, R. C. (Eds.) (1999), *Accommodation management: Perspectives for the international hotel industry.* London: Thomson.

Wood, R. C. (Ed.) (2000), *Strategic questions in food and beverage management.* Oxford: Butterworth-Heinemann.

9

Managing Facilities

Learning Objectives

By the end of the chapter the reader should be able to:

- Reflect on the issues affecting management of accommodation on board
- Discuss yield management in context
- Examine the systems approach for managing accommodation
- Understand factors affecting quality and operations

This chapter deals with the effective management of accommodation on board, including cabins, public areas, crew areas, and deck areas. In examining this department, a number of issues are considered, including administration, yield management, design aspects, routines and schedules, and environmental issues (Adamo, 1999). The constraints on planning accommodation that were discussed earlier, namely, the need to use space economically, the increasing demand for enhanced passenger facilities, and the current trend to ensure that cruise ships sail at full capacity, are all concerns for operational managers in this department.

When considering space management, problems can arise for this department, with passengers embarking with large quantities of luggage, moving the luggage from shore to ship to cabin and back again and storing personal belongings within the cabin. Passengers make comparisons regarding their cabin/stateroom accommodation on board and hotel accommodation ashore. There is an expectation that the look and feel of the accommodation will be appropriate for the style of cruise, the image presented by the brand, the price paid, the advertised product offer, and similar shore-based products. While passengers acknowledge the space constraint, as well as recognizing that the attractions on board mean that proportionately little time may be spent in the cabin/stateroom, the expectation regarding the accommodation can be high. The lack of alternative accommodation space presents problems for managers faced with critical events that may necessitate temporary or permanent reallocation of accommodation.

Revenue or Yield Management

Revenue management (RM) is the practice of offering the correct type of inventory (cabins/staterooms) at the appropriate price to maximize revenue (Yeoman and Ingold, 1997). RM is a practice that is undertaken in a number of industries to ensure that time and selling strategy are managed effectively to generate the best "yield" (Donaghy et al., 1997). Yield is the function of both the price the cruise company charges for differentiated service options (pricing) and the number of cabins or staterooms sold at each price (seat inventory control). The perishability of the inventory, whether that inventory is cabins on cruise ships, seats on aircraft, hotel rooms, or theater tickets, drives the RM policy and drives the company to optimize profitability.

RM includes the construction and implementation of policies related to the formulation and alignment of price, product, and buyer that will lead to profitability. In this way RM uses predictions regarding inventory and market segments and optimum pricing to create an increase in net yield. Typically, RM is applied when service organizations have a fixed capacity (such as with a cruise ship or a fleet of ships) and when success or failure is dependent on how this capacity is used. These organizations frequently have high fixed costs that are covered when a certain level of sales is achieved. Making an additional sale on top of this break-even point has a marginal impact on costs compared to the impact on revenue.

Costs, Sales, and Markets

The cruise business is investment heavy in terms of the ship itself, the fixtures and fittings, the technical and operational aspects of maintaining the ship, the labor element involved, and the provision of services on board. Once in service, it is difficult to adjust the capacity of a cruise ship, so the critical factor is ensuring that the ship sails on full or as close to full capacity as possible. The costs of adding any extra passengers to reduce unused capacity is relatively, inexpensive, so cruise companies can view actual selling prices with an open mind. It benefits cruise companies to sell inventory as quickly as possible so that they can:

- Gain early access to passengers' money (deposits)
- Be able to ascertain demand at an early stage so as to formulate a robust strategy for maximizing yield
- Make decisions to decrease time uncertainty relating to demand

This means that cruise companies derive advantages from operating initiatives designed to attract early booking. Typical initiatives include time-constrained discounts, free upgrades for early booking; additional incentives for early booking (such as transport to port or onboard credit), and loyalty club membership with access to advance notice with a range of early booking benefits.

In order to make critical decisions about RM, the cruise company needs to be aware of:

- Market segments and consumer buying behavior
- Specific markets to be targeted for specific vacations
- Historical demand and booking patterns
- Pricing knowledge (competitors' rates or rate ranges)

Increasingly, it is becoming possible for cruises to be overbooked. This occurs because, based on analysis of trading patterns, cruise operators make predictions about what is likely to happen with cancellations and no-shows. Overbooking is done to compensate for last moment non-arrivals. The overbooking policy is designed to identify actions, including compensation, that are available if a cruise remains overbooked.

An additional element to be considered by the cruise operator is the multiplier effect. This effect implies that revenue can be generated on board after the booking is made, and therefore the RM system needs to be concerned with a yield that notes the opportunity to attract income through sales on board. This model very much depends on the nature of the "package" on offer. The current trend for vessels over recent years has been to maximize occupancy rates by reducing prices while increasing yield through the combination of volume sales of cruise vacations and revenue generated on board. Sales figures represent the percentage of lower berths that are sold on board. Many operators offer four-berth cabins to single passengers willing to share or to family groups who are travelling together, which, in turn, can increase the occupancy percentage. This practice, in combination with the overbooking policy, means that ships can sail with a stated occupancy level that is greater than 100 percent.

Other features of RM worth considering:

- A complex pricing structure that changes can alienate passengers and cause confusion.
- Passengers in this position may seek alternatives.
- The RM system relies on the yield manager knowing the true availability of inventory.
- The distribution system must be reliable.
- RM requires that the company can forecast accurately (forecasting includes knowledge of customers, booking patterns, no-shows and cancellations, supply factors, and market assessment).
- RM is a strategic decision-making process that needs to take into account alternative scenarios.
- RM is a team activity that use software to analyze complex data.

Administering Accommodation

For the reasons explained above, it is possible to achieve capacity for a vessel that exceeds 100 percent. The process of managing accommodation on board is greatly complicated by inflexibility to deal with accidents or problems on board. Most situations are dealt with at the purser's desk or reception. To be able to deal with these problems, the ship's manager/officer needs to be able to make a judgment about the nature of the case at hand and the options available for solving the problem and be able to make a choice from these options about the solution that would be most beneficial for both the passenger and the cruise company.

In the first instance, the purser's department requires accurate information about cabins and passengers. This database, supplied from the sales office ashore, is crucial because it provides intelligence relating to the passenger that may inform the manager further about the background for any potential problem. The database may also inform heads of department about any surplus inventory that can be used in the event that a change of cabin is unavoidable. Second, the purser's office needs to be aware of the policy for dealing with problems so that he or she is appropriately empowered to make decisions and act accordingly. Third, the manager needs to be able to communicate with all parties to access more information when necessary and seek advice when appropriate. A constant channel of communication is maintained between the purser's department and the accommodation manager to facilitate this type of problem solving. The updated information database is critical for administering the accommodation department, to generate accurate passenger accounts or folios, and to produce information about passengers on board for port authorities.

Aesthetics and Ergonomics

The design of cabin or stateroom spaces, public areas, and crew accommodation is undertaken with a view to ensuring that the resultant product:

- Is suitable for the purpose for which it is designed
- Is acceptable to the user in terms of appearance and functionality
- Meets the needs of the user in terms of quality
- Meets the health and safety requirements that are required on board
- Is maintainable and serviceable
- Is congruent with the brand and brand values

The furnishings, fittings, lighting, décor, and quality of air (air conditioning) are, in totality, the product with which passengers and crew interact. The caliber of the linen, the color and texture of the fabrics, the weave of the carpets, the sheen of the wood finish, and the size and feel of the bed are but a small sample of the variables that contribute to the overall design.

The word "aesthetics" refers to the concept of beauty or taste while "ergonomics" is the study of the relationship of people with their environment (Collins, 1987). In this context, it is reasonable to reflect on the balance that can be achieved when designing the interior of a cabin so that it possesses that essential attractiveness that the passenger may desire while remaining intelligently practical for, among other things, resting, sleeping, changing clothes, reading and relaxing (from the passenger's perspective), and cleaning, tidying, and servicing (from the cabin steward's perspective).

Many difficulties can arise from the ergonomic aspects of designing a product that fits all types of people irrespective of their dimensions. In much the same way that an airline is faced with problems if a passenger is too large for a standard seat, the cruise company may have problems if, for example, the passenger has difficulty getting around in a cabin with limited space. However, specially adapted cabins are available for passengers with certain special needs, and most cruise companies are careful to ensure that they welcome rather than discriminate against passengers who may have a special need.

Accommodation Systems

Housekeeping in any accommodation and facility-oriented business is, according to Ball, Jones, Kirk and Lockwood (2003b), fundamental to a successful operation. On a cruise ship, the cabin or stateroom is the most heavily used area and, as a result, the cabin is likely to be critically examined more frequently and in more detail than any other area on the ship. The personnel working in this area have a distinct advantage over those working in equivalent jobs in hotels, because they have a higher profile and greater opportunity to interact with the passenger. For this reason, the aforementioned success is not simply a reflection of the perceived quality of the physical product but also a measure of the service skills of the cabin steward (Figure 9.1).

Ball, Jones, Kirk, and Lockwood (2003b) state that the system for housekeeping ensures that all rooms are clean and serviced appropriately, provides uniforms and laundry services, creates an element of security for guests and their property, and maintains standards of decoration. There is an overlap between the accommodation office and other departments that provide technical support for repairs and servicing. Frequently, galley cleaning is not part of the housekeeping function but is managed internally by galley staff.

Cabin service work requires a lot of attention to detail, and because of the amount of handling and carrying it can be physically demanding. The number of cabins or staterooms that are serviced by

Figure 9.1: A stateroom on the *'Ocean Village'* (courtesy of Graham Busby)

a steward often depends on the type of ship, type of passenger, and type of cabin. On the *'QM2'*, more experienced stewards service the higher-grade cabins. Larger cabins require more time to service and smaller cabins less time. Some ships require stewards to service the cabins and provide room service duties, whereas others do not. Typically, a cabin steward services approximately 12 cabins.

Teamwork in this department is important in ensuring that tasks are completed swiftly and effectively. Cabin servicing may well be an independent activity but, when stewards cooperate and form close working groups, they can support each other and deal with the types of tasks that are better shared. Teams are most effective in dealing with servicing of large areas such as the inside and outside of public areas. The laundry is another important facility for any cruise ship. It functions under the direction of the laundry master and is supported by a team.

Work Schedules and Routines

The tasks involved in serving accommodation fall under different categories:

- Routine daily tasks, which include vacuuming floors, cleaning bathrooms, making beds, tidying surfaces, and changing dirty towels.
- Regular tasks, which include changing linen (sometimes twice per 10-day cruise) and cleaning walls (bulkheads), ceilings (deck heads), windows, and mirrors.
- Periodic tasks, which include deep cleaning, and shampooing carpets and soft furnishings.

Established routines create a balanced workload that is fair and equitable to the personnel involved. Routines are also established to ensure that all regular and periodic tasks are done to the desired standard. According to Adamo (1999), the maintenance procedures for accommodation can be identified as:

- Non-Routine Maintenance (NRM), such as dealing with leaking taps and bulb blowouts.
- Emergency Response Maintenance (ERM), such as responding to leaks and heating problems.
- Cyclical Planned Maintenance (CPM), the cleaning or servicing items on a regular basis.
- Preventive Planned Maintenance (PPM), which includes inspections and planning for maintenance.

Standard operational procedures (SOPs) are communicated from the cruise company head office to ships in order to establish the standardized model on board. SOPs can take the form of photographs

Figure 9.2: Suite on the *'Arcadia'*

of standard layouts, cabin configurations, and vanity tray setups, supported by text that describes details. The SOPs are prescriptive and communicated from managers to supervisors to staff.

Environmental Issues

Ecological concerns present a serious issue for facilities management. Adamo (1999) identifies the typical problems faced by companies aiming to meet an environmental agenda within a consumer-led society that appears to ignore the ecological impacts connected to the excesses of consumption. The hospitality industry is a volume user of water, energy, consumer goods in general, and scarce luxury items, and the cruise industry can be included in this statement. In addition, the cruise industry has to cope with managing waste products to meet merchant shipping regulations and providing a duty of care to the environment.

Most cruise brands employ an environment officer who reports directly to the ship's master to address environmental matters. Davies and Cahill (2000) describe two drivers for environmental friendliness. Those referred to as "upstream" impacts relate to the influence that can be placed by cruise companies on their suppliers to make sure that supplies meet appropriate environmental criteria, while "downstream" impacts are more related to the education of customers and clients. The authors suggest both are within the remit of the cruise industry.

Case Study: Managing Accommodation on a Grand Class Ship

Edward Green has worked in the cruise industry for 18 years. While he has been a Princess Cruises employee for most of this time, in the past he has also worked for other cruise companies. As the accommodation manager, he reports to the staff first purser (Admin.) and then the passenger services director. His rank is equivalent to a senior assistant purser (two stripes), although this aspect is underplayed because Princess consciously chooses to emphasize the hotel or resort aspect of the vacation rather than the maritime element, with its almost pseudomilitary connotations.

The vessel weighs in at 109,000 GRTs and carries 2,600 passengers when full. In order to maintain the high standards expected on board, Edward's team is both diverse and large. The statistics are as follows:

Total employees: 188
Accommodation manager: 1
Supervisors: 9
Administration assistant: 1
Deck supervisors: 3
Stateroom stewards: 72
Inside public area supervisor: 1
Outside public area supervisor: 1
Inside public area accommodation attendants (ACATs): 38
Outside public area accommodation attendants (ACATs): 12
Utility cleaners: 27
Bell box supervisor: 1
Bell box staff: 7
Laundry supervisor: 2
Laundry staff: 23
Number of cabins serviced by stateroom stewards: 18 to 19
Nationalities accommodation personnel: approximately 50% Filipino, 25% Eastern European (Hungary, Slovakia, Poland, and Romania), 25% Thai, and a few Portuguese and Mexican.
Duration of contracts: Accommodation manager, 6 months; Filipino, 10 months; Thai, 10 months; Mexican, 9 months; and Portuguese, 7 months.

Number of bell boxes (room service): 1
Number of pantries: 30

The department operates around a set of routines and procedures that are to be completed to standards as are laid out by the head office:

- Stateroom stewards servicing cabins, suites, and minisuites look after 18 to 19 staterooms each, which might be a mix of suites and cabins. Although suites take longer to service than cabins, the allocation is rotated to make sure that the balance of work is fair.
- ACATs ensure that public rooms and toilets are serviced. In addition to routine tasks, they undertake scheduled deep cleaning and shampooing of furniture, carpets, and drapes in public areas. They also deal with accidents that result in soiled carpets or soft furnishings.
- ACATs (pool boys) service passenger areas out on decks, looking after open deck furniture and any cleaning duties that are required.
- Utility cleaners service crew areas such as alleyways, bulkheads, crew mess and officers' mess, and they act as officers' stewards.
- Bell box room service teams work from the bell box and deck pantries to deliver room service to cabins, suites, and minisuites.

The job of managing accommodation is demanding, but because of the good working relationships both onboard and ashore problems are few and far between. On each cruise there are meetings to examine performance and to consider emerging issues. The brand head office in Santa Clarita keeps in regular and close contact with managers on board to advise about changes or updates to policy and procedures. Systems are in place for a range of eventualities. For passengers with special requests, lists are generated to ensure that the request is distributed to the correct department and personnel for appropriate action. For cruises where passenger profiles are different from the norm, patterns are identified to anticipate needs such as a greater demand for cribs and high chairs. A reporting mechanism is in place to ensure that defects are identified, reported, and appropriately dealt with. Quality control is practiced by all personnel and, the accommodation manager undertakes regular checks that involve random sampling.

Increasingly, passengers report allergies or special needs, and related plans are put in place to meet passenger requirements. Smoking nor non-smoking cabins on board are not specifically designated, so a passenger stating an allergy to cigarette smoke would be allocated a cabin that had been deep cleaned to remove potential problems. For most general deck areas and public rooms, passengers on the port side (left) may smoke, while the starboard side (right) is nonsmoking. Smoking is not allowed in certain lounges and outside decks where there is a potential hazard.

The accommodation manager is responsible for all aspects of managing the department. The department relies on good quality supervisors who have operational experience and strong interpersonal skills. As Edward states, when commenting upon the management skills of his supervisors, "you don't want supervisors who scream and bawl!" In a growing fleet, getting the right supervisors and managers in place can be a challenge, although experience has shown that the *'Star Princess',* as one of many new large ships joining the Princess brand in a relatively short time, has reached the required operational standard in a very short time. The impetus on the head office to source and then schedule key staff is crucial in achieving this aim. Much of the time, training on board is continuous with, personnel learning on the job and being coached by more experienced personnel. Promotion tends to be from within, thus continually building a model of competence based on shared knowledge.

Interesting problems do emerge from time to time, like the very tall passenger who did not fit into his standard 6-foot, 6-inch bed. In this situation, a bed extender was built on short notice by the joiner. Normally the ship would get advance warning about this and similar problems. There are cabins on board that comply with the ADA (Americans with Disabilities Act), with doors that allow easy access for a wheelchair and ramps to get onto the balcony.

Norwalk-like virus (NLV) has been around for a long time and the company has produced an in-depth handbook of guidelines that are regularly updated. Thus, if any suspected incident is reported, it is dealt with as a matter of routine. For example, if the doctor advises that a junior assistant purser has symptoms (not full-blown), the accommodation manager would contact the crew supervisor, who has a trained "hit squad" of 15 people on 24-hour availability to sanitize that person's cabin. The procedure involves isolating the person with the reported problem so as to contain any potential risk. Fleet cabins, which are crew cabins that are allocated for training purposes, are used to cope with the problem to ensure that the ill person is not in contact with others.

On arrival and departure days the accommodation department manages baggage on board. Around 129 baggage cages are filled with the departing passengers' luggage and taken ashore. Conversely, the joining passengers' luggage is boarded using a similar system. ACATs, utility cleaners, and bell box staff handle the baggage, collecting, distributing, and offloading as required. Stateroom stewards are not involved in this process. A turnaround in Venice or Barcelona is easier to manage because the ship has a three-day layover. The Caribbean, with a one-day turnaround, is more difficult, especially on two-week cruises when passengers take more luggage.

In Edward's opinion, the US Port Health inspection is not really a problem because onboard standards are very high. The biggest area for this inspection is the galley, but for his department Port Health officers look at pantries, onboard cleaning routines for dishwashers, and the logs and records related to these areas. They will check that hot tubs are sanitized once every seven days, that potable water is monitored and in perfect condition, and that swimming pools are thoroughly tested. Swimming pools and hot tubs are tested by a pool boy every four hours to take readings, copies of which are taken to the engine control room. The pool boy highlights any discrepancies but engineers will deal with the problem. General housekeeping does not hold dangerous chemicals. The ship's environmental officer, who reports directly to the captain, monitors all potential risk areas from an environmental safety or compliance point of view.

Edward is very aware that soft skills are vital when managing accommodation services. Being able to construct and motivate a team, treating people with respect, and understanding cultural differences are each important elements in the complex task of managing a working community at sea. For example, Edward notes that some nationalities seem more aggressive; they are not necessarily aggressive but they come across as such because of how they do things. People who work on a cruise ship need to get along with both crew and passengers. He says it can be a long time for some to be away from home, and over the years the routines have changed. For example, on the *'Island Princess'*, stateroom stewards used to service 12 cabins with room service—now they service 19 cabins with balconies. Edward also suggests that over the years he has been at sea, passengers have become more demanding, influenced partly by demographic changes and depending on the time of year and the cruising season. Also, as cruise companies have merged, cost control and budgets have been managed more effectively.

Princess Cruises operates a passenger service credo called CRUISE, which has evolved over the eight years since it was introduced. The CRUISE program stands for Courtesy, Respect, Unfailing In Service Excellence. It is intended to create a culture of friendly care within an environment where the crew strives for the best level of service. Edward wholeheartedly supports the strategy but highlights those repeat passengers who know the credo and who place additional burdens on the cruise ship personnel. In particular, when the ship is full—which is becoming more frequent—cabin moves are a problem.

Case Study Questions

1. How is quality addressed in the accommodation department?
2. What are the critical issues to consider when managing this department?

Summary and Conclusion

In the conclusion to the last chapter, the role of accommodation manager was introduced as being parallel to that of the food and beverage manager. There are many similarities, such as the demand for high-volume, sometimes 24-hour service, the highly visible nature of the job, the high expectation held by passengers, and the need for systematic operations. There are also unique aspects, such as the points during the day and during the cruise when there are heavy demands and lighter demands. Turnaround day is always hectic.

This chapter has highlighted key issues connected with accommodation, such as revenue management and how it affects sales, administration and the link with accommodation management, the part played by aesthetics and ergonomics, and the importance of environmental concerns. Some of these points reemerge in the next chapter, which addresses health, safety, and security.

Glossary

Credo: System of principles or beliefs.
Inventory: Totality of (for example) cabins available for sale.
Port: Left side of a ship.
Profit: Net income; the excess of revenue over outlays in a given period of time.
Revenue: The entire amount of income before any deductions are made.
Starboard: Right side of the ship.
Turnaround: Indicates a home port or a port where the cruise starts and finishes.
Yield: The quantity of something (for example, rooms) that is sold in a particular time period.

Chapter Review Questions

1. What is yield management or revenue management?
2. Define ergonomics and aesthetics and describe how these affect a cruise ship.
3. How can scheduling in the accommodation department be managed?
4. How are environmental issues addressed on board?

Additional Sources of Information

http://www.cruisecritic.com/—cruise critic
http://www.cruiseindustrynews.com/—cruise industry news
http://www.cruise-reports.com/Subscribers/memberlogin.htm—cruise reports
http://www.cruisingnews.com.au/default.asp—cruising news
http://www.prowsedge.com/—Prowse Edge—cruising information

References

Adamo, A. (1999), Hotel engineering and maintenance. In C. S. Verginis & R. C. Wood (Eds.), *Accommodation management: Perspectives industry.* London: Thomson.

Ball, S., Jones, P., Kirk, D., and Lockwood, A. (2003), *Hospitality operations: A system approach*. London: Continuum.

Collins (1987), Concise English Dictionary. London: Guild Publishing.

Davies, T., and Cahill, S. (2000), *Environmental implications of the tourism industry.* Retrieved 17 January 2005, from http://wwww.eldis.org/static/DOC10089.htm

Donaghy, K., McMahan-Beattie, U., and McDowell, D. (1997), Yield management practices. In I. Yeoman & A. Ingold (Eds.), *Yield management: Strategies for the service industries.* London: Cassell.

Yeoman, I., and Ingold, A. (Eds.) (1997), *Yield management: Strategies for the service industries.* London: Cassell.

10

Health, Safety, and Security

Learning Objectives

By the end of the chapter the reader should be able to:

- Appreciate the range of issues that can affect health, safety, and security for people on cruise ships
- Appreciate the implications and outline components of the US Public Health Service vessel sanitation program (VSP) operated by the Centers for Disease Control and Protection (CDC)
- Understand aspects of food hygiene that have importance for cruise ship operators
- Reflect on the regulatory framework and the implications for cruise operators in relation to port health authorities
- Consider the implications of the introduction of the International Ship and Port Facility Security (ISPS) Code
- Understand the issues relating to providing a service to customers who have special needs

The image of cruising reflects what many would perceive as a leisure focused utopian dream. However, as with most travel, there are distinct and diverse perils that can exist and that demand attention. This chapter is designed to help the reader appreciate the range of issues that can potentially affect health, safety, and security for people on cruise ships. These complex subjects pose serious challenges for society, and the ramifications for many industries involved in travel, tourism, and leisure have been severe. In the first instance, medical cases concerning the health of passengers are highly visible and attract widespread media attention. The norovirus, also referred to as the Norwalk-like virus, can be a concern, and organizations such as US Port Health have been active in working with the cruise industry and other industries to address the problems that can arise. Although highly visible in the media, this health issue is not the only medical problem that can be faced by cruise operators.

The US Public Health Service vessel sanitation program (VSP) operated by the Centers for Disease Control and Protection (CDC) plays a significant role in helping to make cruise ships safe and hygienic (US Public Health Service, 2005b). This organization promotes good practice, provides information and training, and identifies potential hazards that could lead to the emergence of risk for passengers and crew.

Security on board is paramount, and the cruise industry has expanded and become more successful by presenting itself as a secure option for a vacation. The International Maritime Organization (IMO) has taken the lead in aiming to provide an international framework to ensure that safety and security remain center stage. The introduction of the International Ship and Port Facility Security (ISPS) code was a reaction to the heightened tensions relating to potential threats to shipping in general. This chapter considers key health, safety, and security issues concerning destinations for those who work in this industry. Finally, the chapter concludes by reflecting on the provision of services to customers who have special needs.

Health, safety, and security are complex subjects, and it is not possible to cover every aspect of this subject in detail in this particular section. Readers are therefore prompted to undertake further study to develop their understanding of these areas and to make sure that they remain vigilant about keeping their learning in line with current best practice. This chapter deals with the issue of security in general terms. Much of what is done in securing a vessel is understandably confidential, and comments made in this section describe plans and actions that will not compromise this confidentiality.

Centers for Disease Control Vessel Sanitation Program

The US Public Health Service's Centers for Disease Control and Protection (CDC) introduced the vessel sanitation program (VSP) in the early 1970s because of several disease outbreaks on cruise ships. The VSP primarily targets gastrointestinal illnesses. Over the years, the relationship with the cruise industry has matured, and although the CDC is still seen as a powerful agency in terms of control and regulation, it has evolved to provide assistance and training in order to achieve best practice (US Public Health Service, 2005b).

The CDC is best known for its sanitation inspections that score cruise ships on a 100-point scale. Vessels that fail to score 86 or more points are deemed to have failed the inspection. Ships that score an 85 or lower are therefore declared to have an unsatisfactory sanitation level and have to be reinspected, usually within 30–45 days, to determine if conditions have improved. The CDC asserts that a ship with a lower score is likely to have a lower general level of sanitation, although that does not automatically imply an imminent risk for gastrointestinal illness (GI). The CDC declares that since the program began, the number of disease outbreaks on ships has declined, even though the number of ships sailing and the number of passengers carried has increased significantly.

The VSP undertakes targeted surveys to identify problems relating to GI. This attention is triggered if the total number of passengers or crew members with such an illness reaches two percent of the total passengers or crew members onboard. An investigation may also be undertaken if an unusual pattern or characteristic associated with GI is found. To achieve this, ships are required to maintain a log of passengers and crew who have reported symptoms for GI and who may have requested medicines to treat diarrhea. If a case emerged from the data that caused concern, the VSP would analyze the risk associated to any outbreak, review practices on board, aim to identify infectious agents, develop a prevention and control response, and evaluate the implemented response. The VSP acts to identify the problem, deal with the problem, and make sure the problem does not re-emerge.

The CDC also assists when a newly built or upgraded vessel is being planned and constructed. This ensures the vessel is in an optimum position to meet public health requirements and can help with issues such as the correct location of hand-washing facilities, the correct way to design and construct storage and preparation areas for food and drink (including potable water supplies), and temperature control management for storing food. In its facilities in Florida, the organization offers training for cruise staff on recommended standards, the reasons for the standards, and how to comply with the standards. The following points are covered by their training programs: water storage, distribution, protection and disinfection; food protection during storage, preparation, cooking, and service; employee practices and personal hygiene; general cleanliness, facility repair, and vector control (a vector is any insect or arthropod, rodent, or other animal of public health significance capable of harboring or transmitting the causative agents of disease to humans); and the potential for contamination of food and water.

The Norovirus

The norovirus was previously known as the Norwalk-like virus (Anon, 2002) after a virus that was first identified in 1972 following an outbreak of gastrointestinal illness in Norwalk, Ohio. Norovirus is a collective name for a group of viruses that can affect the stomach and intestines. In some cases, these viruses can cause gastroenteritis, an inflammation of the stomach and large intestine. While gastroenteritis is frequently known as a calicivirus infection or a form of food poisoning, it may not always be related to food. Also, norovirus is sometimes called "stomach flu," although it is not related to the flu, or the "24-hour stomach bug."

People who contract a norovirus tend to present symptoms such as vomiting, diarrhea, and stomach cramps (Widdowson et al., 2004). Children may vomit more than adults. In some cases people may also develop low-grade fever, chills, headache, muscle aches, nausea, or fatigue. The illness can begin quite rapidly, and the infected person may feel very sick. The illness tends to last for one or two days. Noroviruses are found in the stool or vomit of infected people. They can also be present on surfaces that have been touched by people who are infected. Critically, outbreaks have occurred more often where relatively large numbers of people are contained in a small area, such as trains, buses, schools, army barracks, restaurants, hospitals, nursing homes, catered events, and cruise ships (Cramer et al., 2003). The illness can be incubated for 24–48 hours and last 12–60 hours.

Noroviruses are most readily associated with cruise ships (Ramilo et al., 2004) but this perception is incorrect. There are many more cases ashore than at sea. The reporting mechanism connected to the VSP and the work of health officials who track illnesses on cruise ships enable shipboard problems to be identified and dealt with more effectively than they are ashore. Often the problem emerges because the illness is carried on board and then subsequently spread by passengers. The layout of a cruise ship, with accommodation configured in close proximity, is thought to contribute to the amount of interaction and person-to-person contact.

People can become infected with the virus in a number of ways, as shown in the list below. It is important to note that the virus is more likely to originate ashore than on board a cruise ship and to identify that this is a people problem not a cruise problem—although of course the industry must deal with the consequences.

- The virus can infect food and drink. In particular, ready-to-eat foods such as shellfish, deli foods, sandwiches, dips, salads, peeled fruits, and communal foods that require handling can be affected. The food may have been contaminated before purchase. Contaminated water can also be a threat because of inferior sewage treatment of contamination in pools, rivers, swimming pools, well water, or ice.
- A person can inadvertently touch a surface or object that is infected with noroviruses and then touch her or his own mouth, nose, or eyes.
- One-to-one contact with a norovirus-infected person can occur by being in the immediate vicinity while someone is vomiting, which suggests that there is an airborne risk. It can also occur when caring for an infected person or using the same utensil to share food with an infected person. It can even occur when shaking hands (which explains why some staff on cruise ships have been known to replace the ritual of shaking hands with touching elbows).
- Not washing hands after using the bathroom or changing diapers and before eating or preparing food is also a risk for infection.

Noroviruses are highly contagious but are not generally regarded as serious (Lindesmith et al., 2003). The symptoms are uncomfortable and can be distressing, but there are usually no long-term adverse health effects. Anyone who contracts the virus is advised to contact the doctor or medical staff, to drink fluids (because of dehydration from vomiting or the effects of diarrhea), and to be particularly careful to wash hands frequently. Practice on board is sensitive to outbreaks (Sternstein, 2003). Crew members and passengers are advised to wash their hands often. This should be done after using the toilet, after sneezing or coughing, after changing a baby's diaper, and before eating, drinking, preparing food, or smoking. Frequency of hand washing should increase if someone is unwell. Passengers and crew members are advised to wash their hands with soap and water for at least 20 seconds before rinsing them thoroughly. They are also advised not to touch their mouths because of the risk of infection. An alcohol-based hand sanitizer can be used along with hand washing (US Public Health Service, 2005a). A dispenser is usually located before entering a buffet service area or restaurant.

Regimes on board cruise ships are directed toward prevention, surveillance, and response. The onboard plan is based on isolation, containment, disinfection, investigation, and information/education.

Isolation: This is interpreted as confining the infected person to quarters for three days after the symptoms have ended. Care is recommended to locate the person away from other people who may share the accommodation. Full instructions about personal hygiene should be provided to the infected person.

Containment: The area that may be affected should be dealt with by a specially trained, equipped, and prepared hit squad. Access to the area should be carefully restricted. Infected people should be treated by medical or support staff who wear universal precaution garb (gown, gloves, and mask). It is recommended that passengers are not charged for this care.

Disinfection: Disinfectants such as accelerated hydrogen peroxide, Virkon, Ecotru, Mikro-Bac II, Mikro-Bac 3, Cryocite 20, and bleach can be used to eradicate the virus in a specific location. Areas and objects that are likely to receive a significant amount of touching by hand should be targeted—railings, banisters, handles, pens, pencils, tables, counters, chips in the casino. The list is endless. Indoor and outdoor facilities and all public areas such as lounges, bars, toilets, buffets, and restaurants may be affected.

Investigation: A full history should be taken to identify potential causes.

Information/education: This process involves informing the crew and passengers about any outbreak and telling them what it is and what it means. Give advice about how to deal with the situation including reporting problems and taking precautions. The crew should be fully trained to understand the issues before commencing work, either through an induction event or a training program.

Vessel Sanitation Program Inspection

The VSP calls for all cruise ships with a foreign itinerary calling into the United States and carrying more than 13 people to be inspected twice yearly by a team of environmental health officers. Inspection details are described in the VSP manual (US Public Health Service, 2005b). The VSP operations manual was revised in 2005 to take account of new technology, advances in food hygiene, and emerging biological agents that can cause disease to humans. The manual is used to guide and educate cruise ship operators and crew and to focus the CDC in its inspection routines. The following points are summarized from the manual.

Water: Potable water refers to drinking water. The process of pumping water from shore to ship is known as bunkering (this can also be the term used when fuel and stores are taken on board). Drinking water is expected to meet World Health Organization (WHO) standards. It should be regularly sampled (every 30 days or less) and tested to provide a microbiological report to confirm that it meets expected standards. The ship retains records on testing for 12 months. Generally, water cannot be produced on board (through reverse osmosis, distillation, or another process) when the ship is at anchor, in polluted areas, or in harbor. The manual provides full technical information relating to water and water systems on board.

Swimming pools: Flow-through sea water pools may be used only when the ship is under way and at sea beyond 12 km from land. The pool is drained prior to arrival in port and remains empty while in port. In some circumstances a pool may be left full as long as appropriate procedures are in place to disconnect the filling system and to provide appropriate filtration and halogenation. Water safety tests must meet desired levels before bathers can use the pool. Recirculating pools must be effectively filtered in accordance with the filtration manufacturer's instructions. Water quality must be monitored and must exceed expected minimum standards.

Whirlpool water is filtered, and the filters are inspected regularly and changed every six months. Water is changed daily. Safety signs and depth markings for pools must be displayed prominently. Temperature controls must prevent the water from exceeding 40 degrees centigrade. Safety equipment must be provided as directed. Babies or young children in diapers and children who are not toilet trained are not permitted in the pool.

Food Safety: The person in charge of food production and food safety on board must possess an appropriate level of knowledge about food-borne disease prevention, Hazard Analysis Critical Control Point (HACCP) principles, and the VSP food safety guidelines. Appropriate certification from the United States or overseas is acceptable evidence, although evidence is also demonstrated through practice observed on board and the ability to answer questions during the inspection (see Table 10.1).

In this context knowledge encompasses personal hygiene and how it affects prevention of food-borne disease; the areas of responsibility held by a manager in charge of a food production team; the symptoms associated with food-borne diseases; the importance of holding time and temperature control for potentially hazardous food; the hazards related to raw or undercooked eggs, meat, poultry, and fish; safe cooking times for potentially hazardous food; management and control concerning cross-contamination, hand contact with ready-to-eat foods, hand washing, and general hygiene; food safety; provision of equipment in appropriate condition; procedures for cleaning and sanitizing; storage, usage, and handling of toxic or poisonous materials; and operational management and control of production and to service routines to prevent problems associated with food hygiene and safety.

Managers are expected to ensure that their staff adopt safe and hygienic practices as laid down by the manual. In general terms all food must be safe, unadulterated, and sourced appropriately to meet guidelines. Potentially hazardous foods must be received at a temperature of 7°C or below. Food is protected from contamination by being stored in a clean, dry location, 15 cm above the deck in a location where it will not be contaminated. Food may not be stored in places such as locker rooms, toilets, dressing rooms, garbage rooms, mechanical rooms, and open stairwells. Food on display must be protected from contamination.

Table 10.1: **Guidelines for food safety (US Public Health Service, 2005b:51)**

The person in charge of the food operations on the vessel shall ensure that:

1. Food operations are not conducted in a room used as living or sleeping quarters;
2. Persons unnecessary to the food operation are not allowed in the food preparation, food storage, or warewashing areas, except that brief visits and tours may be authorized if steps are taken to ensure that exposed food; clean equipment, utensils, and linens; and unwrapped single-service and single-use articles are protected from contamination;
3. Employees and other persons such as delivery and maintenance persons and pesticide applicators entering the food preparation, food storage, and warewashing areas comply with the guidelines in this manual;
4. Food employees are effectively cleaning their hands, by routinely monitoring the employees' handwashing;
5. Employees are observing foods as they are received to determine that they are from approved sources, delivered at the required temperatures, protected from contamination, unadulterated, and accurately presented, by routinely monitoring the employees' observations and periodically evaluating foods upon their receipt;
6. Employees are properly cooking potentially hazardous food, being particularly careful in cooking foods known to cause severe food-borne illness and death, such as eggs and comminuted meats, through daily oversight of the employees' routine monitoring of the cooking temperatures using appropriate temperature measuring devices properly scaled and calibrated;
7. Employees are using proper methods to rapidly cool potentially hazardous foods that are not held hot or are not for consumption within 4 hours, through daily oversight of the employees' routine monitoring of food temperatures during cooling;
8. Consumers who order raw or partially cooked ready-to-eat foods of animal origin are informed that the food is not cooked sufficiently to ensure its safety;
9. Employees are properly sanitizing cleaned multiuse equipment and utensils before they are reused, through routine monitoring of solution temperature and exposure time for hot water sanitizing, and chemical concentration, pH, temperature, and exposure time for chemical sanitizing;
10. Consumers are notified that clean tableware is to be used when they return to self-service areas such as salad bars and buffets;
11. Employees are preventing cross-contamination of ready-to-eat food with bare hands by properly using suitable utensils such as deli tissue, spatulas, tongs, single-use gloves, or dispensing equipment; and
12. Employees are properly trained in food safety as it relates to their assigned duties.

Table 10.2: **The seven principles of HACCP (Food and Drug Administration, 2001)**

1. Analyze hazards. Potential hazards associated with a food and measures to control those hazards are identified. The hazard could be biological, such as a microbe; chemical, such as a toxin; or physical, such as ground glass or metal fragments.
2. Identify critical control points. These are points in a food's production—from its raw state through processing and shipping to consumption by the consumer—at which the potential hazard can be controlled or eliminated. Examples are cooking, cooling, packaging, and metal detection.
3. Establish preventive measures with critical limits for each control point. For cooked food, for example, this might include setting the minimum cooking temperature and time required to ensure the elimination of any harmful microbes.
4. Establish procedures to monitor the critical control points. Such procedures might include determining how and by whom cooking time and temperature should be monitored.
5. Establish corrective actions to be taken when monitoring shows that a critical limit has not been met—for example, reprocessing or disposing of food if the minimum cooking temperature is not met.
6. Establish procedures to verify that the system is working properly—for example, testing time-and-temperature recording devices to verify that a cooking unit is working properly.
7. Establish effective record keeping to document the HACCP system. This would include records of hazards and their control methods, the monitoring of safety requirements and action taken to correct potential problems. Each of these principles must be backed by sound scientific knowledge, for example, published microbiological studies on time and temperature factors for controlling food-borne pathogens.

Cooking times:

- A temperature for raw eggs of 63°C or above must be held for a minimum of 15 seconds.
- A temperature for ratites (ostrich, emu, and rhea) and injected meats of 68°C or above must be held for a minimum of 15 seconds.
- A temperature for poultry and wild game of 74°C or above must be held for a minimum of 15 seconds.
- A temperature for whole roasts of 63°C or above must be held for a minimum of 15 seconds (note that there are alternative conditions for rare meat cooking and for preparation of other foods noted in the VSP manual).

Food cooling: Potentially hazardous food shall be cooled within 2 hours from 60°C to 21°C and within 4 hours from 21°C to 5°C or less.

Holding temperatures: Potentially hazardous food shall be held at 60°C or above (except roasts, which may be held at 54°C or above) or at 5°C or less.

Full details about food safety and hygiene relating to food storage, handling, service, equipment and equipment care, cleaning, sanitizing, and managing the vessel, the galley, and associated areas found in the VSP operations manual (US Public Health Service, 2005b). It is possible for operators to apply to the CDC for variances to requirements. These variances may be granted if evidence is provided to show that the request is reasonable and will not jeopardize crew and passenger health.

There are a number of non-US ports that, when visited, are likely to prompt a visit from the National Port Health authorities, such as Southampton or Sydney. In the main, the requirements of these authorities overlap those that are operated by the VSP.

The process of inspection identifies that it is good practice to adopt a Hazard Analysis and Critical Control Point (HACCP) approach for food and drink. This approach involves seven principles (see Table 10.2).

Safety at Sea

For a significant number of people, the act of terrorism and the threat of terrorism present major concerns. Since 9/11, the apparent safety and relative peace enjoyed by those in the wealthier nations of the world has been compromised as governments have taken steps to challenge those who harbored terrorists or condoned terrorism. The corollary to this situation has been increased vigilance at borders and increased security in general. The security implications have actually helped the cruise industry, which can capitalize on flexible itinerary planning and the opportunity to move in and out of world regions depending on risk (Parker, 2004). The downside has created more bureaucracy, longer lines for passengers and crew at security desks, higher levels of intrusion into individuals' lives, increased costs, and greater complexity when planning. Cruise passengers are said to have accepted the implications of the heightened security as a normal part of that type of vacation under the current circumstances (Scorza, 2004).

Under the watchful eye of the IMO, shipping has operated within a framework that raises the bar for the safe operation of vessels. SOLAS and MARSEC are key elements of marine safety and security (see Chapter 3). In December 2002, a conference attended by 108 contracting governments to the 1974 SOLAS Convention, observers from two IMO Member States, and observers from the two IMO Associate Members was held (IMO, 2002). The attending representatives agreed to a series of measures intended to strengthen marine security and aimed at preventing and suppressing terrorist acts against shipping. The resulting code is an amendment to the SOLAS agreement and is known as the International Ship and Port Facility Security (ISPS). ISPS provides detailed security-related requirements for governments, port authorities, and shipping companies in a mandatory section, together with a series of guidelines about how to meet these requirements in a second, nonmandatory section (see Table 10.3).

The target date for implementing ISPS was July 1, 2004, and although some contracting states did not achieve compliance, the majority have (NSnet, 2004). The implications for a ship visiting a port that does not comply can be serious, because the next port of call on the route may view the ship as being contaminated from a security point of view. This could result in raised security measures or even, in a worst case scenario, refusal of entry.

Contracting governments set the risk level that is appropriate for a port facility or for a ship (see Table 10.4). The levels are designed for easy communication of a clear message. The levels correspond to the basic assumption that a hazard with a low probability is a low risk and a hazard with a high probability is a high risk (Smith, 2004).

Both the ship and port facility are responsible for monitoring and controlling access, monitoring the activities of people and cargo, and ensuring that security communications are readily available. The Ship's Security Officer (SSO) is accountable to the ship's master. This notes the master's ultimate

Table 10.3: **ISPS process**

Contracting Government Risk Assessment

1. Identify and evaluate important assets and infrastructures that are critical to the port facility
2. Assessment must identify the actual threats to those critical assets and infrastructure in order to prioritize security measures
3. Assessment must address vulnerability of the port facility by identifying its weaknesses in physical security, structural integrity, protection systems, procedural policies, communications systems, transportation infrastructure, utilities, and other areas within a port facility that may be a likely target
4. The port facility are required to develop port facility security plans, to appoint port facility security officers and to have access to certain security equipment

Company and Ship

1. Company is required to appoint a designated Company Security Officer (CSO)
2. Ship is required to appoint a Ship Security Officer (SSO)
3. CSO to prepare Ship Security Plans for approval
4. The ship is required to have in place ship security plans, to appoint ship security officers and company security officers and to have access to certain onboard equipment

Table 10.4: **Level of risk and action**

Risk and action	*Level one—normal threat*	*Level two—medium threat*	*Level three—high threat*
Port facility	Minimum operational and physical security measures the port facility has established as essential.	The additional, or intensified, security measures the port facility can take to move to when instructed to do so.	The possible preparatory actions the port facility could take to allow prompt response to the instructions that may be issued.
Ship	Minimum operational and physical security measures the CSO has established as essential.	The additional, or intensified, security measures the ship itself can take to move to and operate at.	The possible preparatory actions the ship could take to allow prompt response to instructions that may be issued to the ship.

responsibility for ship safety and security. On cruise ships the SSO manages a team of security professionals who are frequently sourced from the armed forces or police. Ships are required to carry an International Ship Security Certificate and to establish a security alert communication system that can be activated from the bridge and another location on board to identify if and when a serious breach of security is at hand. This alert is to be communicated without sounding any alarm on the ship itself.

Interestingly, not all ports have followed normal convention to sign up to ISPS. In a case reported in Fairplay International Shipping Weekly (2005b), Porto Cervo on the island of Sardinia has become the first Italian port to refuse to invest in security measures that are required for ISPS and effectively has removed itself as a cruise destination. It has been suggested that this decision may have something to do with ensuring that the destination retains exclusivity.

Parker (2004:15), notes that although security associated with cruise ships often relies on technology such as X-ray machines and scanners, higher levels of security are achieved by creating a "security philosophy and mindset" among the crew, staff, and officers. This suggests that a well-trained crew that is aware, observant, and alert to potential security issues is an advantage in the effort to control risk.

Assessing Risk

In the terminology of risk assessment, "a hazard (causal) is a potential threat to humans and their welfare; a risk (likely consequence) is the probability of a hazard occurring and creating loss; disaster (actual consequence) is the realization of a hazard" (Smith, 2004:12). According to Faulkner (2001), tourism disaster management should involve coordination, consultation with all parties, commitment, risk assessment, prioritization, the development of protocols, a capability audit, a command center in the event of a disaster, a media communication strategy, a warning system, and some flexibility. Risk is a term that is frequently evaluated in relation to consequence and likelihood.

The nature of this topic is such that one's attention is immediately drawn to the high impact, high profile events that can dominate world attention for lengthy periods, yet risks may be small by nature or start small and ultimately conclude with a disproportionately damaging outcome. It is useful to make a distinction between a crisis and a disaster in this sense; a crisis is said to be an issue emerging from poor or ineffective planning and management while a disaster is thought to be relatively unavoidable because of natural events (Faulkner, 2001). Incidentally, Smith (2004) draws attention to the notion that a disaster is a social phenomenon. That is, if humans are not involved the critical event is not viewed as a disaster. Planning for risk is common sense but it is also good business sense. The logic of this type of planning can make the difference for a company and can demonstrate the way the company values its clientele and staff, the commitment the company has for its business, and the maturity the company shows in dealing with some of today's potentially chilling realities.

Table 10.5: **Facilities on board by type**

Hotel facilities		*Ship facilities*	
Passenger Facilities	Staterooms/cabins	Comfort system	Air conditioning
	Stairways and halls		Water and sewage
	Public areas		Stores
	Public areas (outdoor)		Engine room
Crew Facilities	Crew cabins	Machinery	Pump room
	Crew messes and bars		Steering and thrusters
	Crew common areas		Fuel and oil
	Crew stairs and corridors		Water and sewage
Task-related Facilities	Tender boats	Tanks/Voids	Ballast and voids
	Stern marina		Lifeboat
	Special attractions		Life raft
Entertainment Facilities	Casino	Safety	Sprinklers
	Swimming pool		Detectors and alarms
	Jacuzzi		Low level lighting
	Cabaret		Life jackets
	Games areas		
	Nightclub		
	Shore excursions office		
Service Facilities	Passenger service		
	Catering production and service areas		
	Hotel service areas		
Others	Shops		
	Beauty salon		
	Nightclub		
	Medical center		
	Photo shop		
	Internet		

The types of hazards that exist can be categorized as natural or environmental (which can include severe storms, earthquakes, and flooding), biological, technological, or what Smith (2004:8) calls "new concern threats," which allude to matters such as terrorism. These hazards can have implications for humans, goods and property, and the environment. In terms of risk, they can expose vulnerability that can test human resilience or responsiveness and also test the reliability of measures that are put in place to address the event (Smith, 2004).

When commenting on risk and risk assessment, Lois, Wang, Wall, and Ruxton (2004) delineate the two types of facilities on a cruise ship: the hotel facilities and ship facilities. The hotel and ship facilities can be disaggregated to identify components as is portrayed in Table 10.5.

In addition, Lois et al. (2004) describe cruise shipping as different from other shipping because the passengers' needs must be accommodated in the following ways: the ship's design and structure (for example, the requirement for appropriate traffic lanes, the division of accommodation for crew and passengers); the appropriateness of docking facilities or support for tendering; the servicing of supply, fuel, and waste management; the itinerary based on passenger demand; the terminal facilities required to process people and provide shoreside facilities and services; and the need to have access to a transport infrastructure for home ports or turnaround ports and destinations. These characteristics all factor into the analysis of risk for a cruise ship.

The IMO has proposed that shipping companies adopt Formal Safety Assessment (FSA). The FSA is described as a structured and systematic approach to risk analysis and as a tool for interpreting the rules and regulations that must be implemented by those responsible for shipping safety.

The methodology aims to create a balance between technical, operational, and human factors and a balance in terms of maritime safety, environmental concerns, and cost factors.

The FSA recommends these five steps of risk analysis:

1. Hazard identification (based on potential and relevant accident scenarios that could occur, together with likely causes and outcomes)
2. Risk assessment (evaluation of risk factors)
3. Options for controlling risk (devising measures to control or reduce the identified risks)
4. Cost-benefit assessment (calculating the cost effectiveness of each risk control option)
5. Establishing recommendations for decision making (making an informed decision taking all the facts into account)

Lois et al. (2004) counsel that hazard identification can emerge in a number of ways, including brainstorming by a team of experts, formal studies of operations based on systematic reflection, analysis of failure mode and effects concentrating on known potential defects or problems, and analysis using a flow chart. The last option can result in a model that recognizes five logical phases for a cruise: the embarkation (passenger arrival, checking in, establishing onboard account, information and key or keycard, and photo opportunity), getting under way (welcoming passengers, directing to cabins, delivering luggage, safety information), the cruise (normal cruising routines, daily program), docking or tendering (shore excursions in destination, transit prior to disembarkation), and disembarkation (managing passengers, managing luggage, coordinating transport). This enables the conduct of a study that recognizes the complex situated factors that can affect a cruise (see Table 10.6).

Using this approach, a matrix can be constructed to quantitatively analyze risk. Note that hazard analysis can be gauged according to frequency using the following five-point scale.

1 = remote
2 = occasional
3 = likely
4 = probable
5 = frequent

In terms of consequence a five-point scale is also used to reflect outcome.

Risk assessment interprets the factors that influence hazards at each level to examine implications when practices such as training of crew, design, maintenance, or communication are changed or improved. This consideration of influences can lead to the creation of "what if" scenarios. In effect, this overview enables a systematic process of assessment of those circumstances, influences, and faults that can lead to an event in order to make a judgment about relative risk. This analysis can be

Table 10.6: **Analyzing risk**

Point on scale	*Outcome*	*Implication*
1	Negligible	No first aid, no delay to voyage, no environmental impact, no cosmetic damage to vessel.
2	Minor	Some first aid, some cosmetic damage, no environmental impact, some delay to vessel.
3	Significant	More treatment than first aid required, vessel damage, some missed voyages, some environmental impact.
4	Critical	Severe injury, major damage to vessel, major environmental damage, cancelled voyage.
5	Catastrophic	Loss of life, loss of vessel, extreme environmental damage, cancelled voyages.

Table 10.7: **Risk assessment and cost-benefit analysis**

	Cause	*Incident*	*Accident*	*Consequence*
Causal chain	Passenger consumes too much alcohol	Passenger starts a fight	Crew or passenger injury	Damage to company reputation Lower staff morale
Interventions by stage	Intervention to remove cause	Intervention before incident	Intervention before accident	Intervention before outcome
Potential interventions	A = Training B = Develop policy for sale of alcohol	A = Security staff in area B = Design of bar area (mirrors line of sight, etc.) C = Early warning protocols	A = Procedure for dealing with difficult and unruly individuals	A = Response plan
Cost-benefit ratings (1—very low to 5—very high) Result = benefit divided by cost	A = Cost 3 medium A = Benefit 5 very high (Result 1.66) B = Cost 2 low B = Benefit 5 very high (result 2.5)	A = Cost 5 very high A = Benefit 4 high (Result 0.8) B = Cost 4 high B = Benefit 4 very high (Result 1)	A = Cost 3 medium A = Benefit 4 high (Result 1.33)	A = Cost 3 medium A = Benefit 4 high (Result 1.33)

undertaken against a PESTLE model (see Chapter 5) or, as Lois et al. (2004) suggest, using commercial, regulatory, technical and social, or environmental contexts.

Causal chains can be useful in identifying a chain of events associated with a hazard and in constructing a list of countermeasures that can avoid or mitigate against the hazard based on interventions that could involve human resources, physical resources, or systems or processes. Risk control options suggest interventions that can remove a cause by focusing on the setting and conditions within the setting, interventions before a prospective incident by considering ways the emerging issue can be identified in timely fashion and an alert raised to prompt action, interventions before an accident such as drills or special incident protocols, and interventions before the outcome such as response plans.

It is also possible to measure a risk in terms of cost and balancing that against the benefit. The cost-benefit assessment technique can address the events that can lead up to and create a hazard and assess them on a five-point scale for benefit and a five-point scale for cost to calculate a result on the basis of benefit divided by cost.

This sequence of analysis is exemplified in Table 10.7, using a hypothetical situation that is unlikely to be encountered on many ships but provides a graphic representation of the process:

Hazard: drunk passenger causes fight in bar. Frequency rating, 2 (occasional, once or twice a month): consequence rating, 2 (possible first aid and damage to bar furniture and glassware).

This structure implies that managers will be in a position to arrive at an informed decision based on this type of analysis, which can take into account complex situated factors and can lead to a prioritization for action in a logical manner. As the table suggests, there are considerable advantages to be accrued in focusing on the causes as a priority in terms of cost benefit.

Providing a Service to Customers Who Have Special Needs

The Americans with Disabilities Act (ADA) is a powerful piece of legislation that protects the rights of US citizens who have a disability and aims to protect them from discrimination (US Department of Justice, 2004). With reference to the cruise industry, the legislation affects any cruise company with a vessel registered in the United States or with a ship that uses US ports. In the UK there is similar legislation in place to protect those who may be discriminated against in this way (UK Government, 2005). It makes good business sense to ensure that provisions are made for all guests with due regard to these laws, to be compliant but also to be seen as a responsible business that cares about its customers.

The Act requires people with disabilities to have equal treatment, equal access, and the removal of any barriers that may inhibit access. For cruise companies this can start with design and construction that take into account the need to ensure that ADA-compliant staterooms are provided and that facilities adhere to the ADA guidelines.

Cruise ships vary in terms of when they were constructed, their target market, the type of itinerary they will be covering, the range and scale of facilities, the carrying capacity, and so on. This implies that passengers should consult carefully with their travel agents to ensure that they are happy with the provision that is being offered. An example of advice provided to passengers by Princess Cruises can be seen in Table 10.8.

Table: 10.8: **Princess access-adapted from Princess Cruises (2005)**

PRINCESS ACCESS

Princess makes every effort to accommodate our passengers with disabilities. Just be sure your travel agent notifies us of your wheelchair usage and/or any other special needs prior to sailing. We have wheelchair-accessible staterooms on all Princess ships. Each ship has a limited number of wheelchairs available to pre-reserve, but please be aware they must be confirmed in advance. Wheelchairs reserved for onboard use cannot be used for pre-and/or postcruise land tours or hotel packages. If you require a mobility device in those instances, you must provide your own. When bringing your own wheelchair, we highly recommend collapsible wheelchairs, as the width of stateroom doors varies. Some Princess ships have areas that are not wheelchair accessible.

Accessibility varies widely on pre- and postcruise land tours and hotel packages. Not all products have lift-equipped transportation options available and, when available, arrangements must be secured in advance to accommodate your needs for pre- or postland tours. To ensure we can accommodate your needs, please call the Tour Quality Department at XXXXXXXXXXX.

If you have purchased a Princess Transfer at the start or end of the cruise, be aware that lift-equipped transportation may be available in your port of embarkation or disembarkation. When available, arrangements must be secured in advance to accommodate your needs. To ensure we can accommodate your needs, we ask that you contact us at XXXXXXXXXXXX.

Passengers utilizing mobility devices with batteries are advised that the batteries must be of a dry cell type and they will be required to be stored and recharged in the individual staterooms. Scooters and or wheelchairs may not be left in hallways. Because we are not staffed with specially trained personnel to assist passengers with physical challenges, we recommend you be accompanied by someone who is physically able to assist you both ashore and on board if necessary. Travelers with disabilities should check in with the onboard Tour Office to ensure all prereserved tours can accommodate their needs. Not all port facilities are easily accessible for those using mobility devices.

Ports of call may be accessed by a variety of methods including, but not limited to, a ramped gangway, series of steps or by tender. In some cases, you may be able to access the tender; however, the shoreside facility is not accessible. With your safety and comfort in mind, the decision to permit or prohibit passengers from going ashore will be made on each occasion by the ship's Captain, and the decision is final. Those ports, which normally utilize tenders to access the shore, are noted on the itinerary. In many ports of call a mechanism known as a stair climber is used to assist passengers up and down the gangway. The stair climber requires passengers transfer to a Princess wheelchair, which is then connected to the stair climber and operated by the ship's personnel. If you cannot transfer or your personal mobility device cannot be easily disembarked due to size or weight, you may be precluded from going ashore.

If you are traveling with a service animal, please be aware Princess requires notice in advance. Entry regulations vary from port to port and there are some ports that prohibit the landing of animals altogether. Passengers are advised to consult the local authorities at each port of call prior to departure for the necessary documentation.

Princess does not provide food for service animals.

Summary and Conclusion

This chapter provides insight into a number of critical issues that affect the cruise industry. In particular, the role of the CDC and the VSP are examined to help develop understanding about the implications and actions that arise from the inspection regime. The norovirus is also discussed to place the risk associated to the problem in context and to reflect on good practice. It is important for future cruise managers to appreciate these elements of health and safety and also to refer to updates relating to the development of good practice. Much is said about security in our modern world. All aspects of our societies are affected, and the cruise industry is no exception. This is discussed in this chapter along with risk assessment and safety at sea. A cruise ship is a complex machine, yet increasingly there are people wishing to cruise who may have disabilities. This creates a dilemma for the cruise operator, who must meet all individual passenger needs despite the constraints imposed by being on board a cruise ship with gangways, tender operations (boat ports), safety and emergency drills, and the implications of weather at sea. This chapter has considered some of the points concerning ADA in this respect.

Glossary

Contagious: Transmitted by direct or indirect contact.
Distillation: The act of purifying liquids through boiling, so that the steam or gaseous vapors condense to a pure liquid.
Gastrointestinal illness: An illness affecting the stomach and the intestines or bowels.
Incubation: The time that elapses between infection and the appearance of symptoms of a disease.
Infection: An incident in which an infectious disease is transmitted.
Microbe: A term for a microorganism, especially one that causes disease.
Protocol: A formal set of rules or standards.
Reverse osmosis: Water treatment process whereby dissolved salts, such as sodium chloride, calcium carbonate, and calcium sulfate may be separated from water by forcing the water through a semipermeable membrane under high pressure.
Sanitation: Maintaining clean, hygienic conditions that help prevent disease.
Sanitizer: A chemical used to destroy unwanted contaminants such as bacteria and viruses.

Chapter Review Questions

1. What is the norovirus?
2. What protocols can be established on board for dealing with an outbreak?
3. What is the ISPS?
4. What is the role of the CDC?
5. What is the implication for a cruise ship visiting the United States that earns a VSP score of 83?
6. Describe the process of undertaking a cost-benefit analysis associated with the FSA.
7. What are the difficulties associated with complying with UK and US legislation for disability discrimination?

Additional Sources of Information

http://www.imo.org: ISPS
http://www.safetyatsea.net/: *Fairplay* magazine publication
http://www2a.cdc.gov/nceh/VSPIRS/vs_pmain.asp: VSP scores

http://www.cdc.gov/nceh/vsp/ConstructionGuidelines/constructionguidelines.htm: VSP vessel construction guidelines
http://www.cdc.gov/nceh/vsp/pub/mmwr/mmwr.htm: VSP morbidity and mortality weekly reports
http://www.cdc.gov/nceh/vsp/pub/Norovirus/Norovirus.htm: information about the norovirus
http://www.cdc.gov/nceh/vsp/pub/Handwashing/HandwashingTips.htm: hand-washing tips
http://www.mcga.gov.uk/c4mca/mcga-guidance-regulation/mcga-ops-ms-home/dops_ms_guidance.htm: Maritime and Coastguard Agency UK security guidelines
http://www.mcga.gov.uk/c4mca/mcga-dops_ms_applications_and_requirements.pdf: ISPS code
http://www.imo.org/home.asp: Marine security
http://www.imo.org/Safety/mainframe.asp?topic_id=351: Formal Safety Assessment

References

Anon (2002), Outbreaks of gastroenteritis associated with noroviruses on cruise ships—United States, 2002. MMWR. *Morbidity and Mortality Weekly Report*, 51(49), 1112–1115.

Anon (2005), ISPS brings the plod of the plebs, *Fairplay International Shipping Weekly*.

Cramer, E. H., Gu, D. X., and Durbin, R. E. (2003), Diarrheal disease on cruise ships, 1990–2000, *American Journal of Preventive Medicine*, 24(3), 227–233.

Faulkner, B. (2001), Towards a framework for tourism disaster management, *Tourism Management*, 22(2), 135–147.

Food and Drug Administration (2001), HACCP: A State-of-the-Art Approach to Food Safety. Retrieved June 2005, from http://www.cfsan.fda.gov/~lrd/bghaccp.html

IMO (2002), Conference of Contracting Governments to the International Convention for the Safety of Life at Sea, 1974: 9–13 December 2002. Retrieved 2005, June, from http://www.imo.org/home.asp

Lindesmith, L., Moe, C., Marionneau, S., Ruvoen, N., Jiang, X., Lindbland, L., et al. (2003), Human susceptibility and resistance to Norwalk virus infection, *Nature Medicine*, 9(5), 548–553.

Lois, P., Wang, J., Wall, A., and Ruxton, T. (2004), Formal safety assessment of cruise ships, *Tourism Management*, 25(1), 93–109.

NSnet (2004), News archive. Retrieved 8 June 2005, from http://www.nsnet.com/archive-1-2004-06.html

Parker, S. (2004), Adopting a security mindset. Lloyd's Cruise International, April/May, 14–15.

Princess Cruises (2005), Cruise answer book. Retrieved June 2005, from http://www.princess.com/onboard/answer/20 05_Cruise_Answer_Book.pdf

Ramilo, P. B., Augenbraun, M., and Hammerschlag, M. R. (2004), Recent Outbreaks on Cruise Ships, *Infections in Medicine*, 21(1), 14–17.

Scorza, A. (2004), Euro cruise shipping: paying the price for peace of mind, *Fairplay International Shipping Weekly*.

Smith, K. (2004), *Environmental hazards* (4th ed.). London: Routledge.

Sternstein, A. (2003), How good is health care on those big cruise lines? *Forbes*, 171(8), 249–251.

UK Government (2005), Changes to the Disability Discrimination Act. Retrieved June 2005, from http://www.disability.gov.uk/law.html

US Department of Justice (2004), A guide to disability rights laws. Retrieved June 2005, from http://www.usdoj.gov/crt/ada/cguide.htm

US Public Health Service (2005a), Noroviruses. Retrieved 14 June 2005, from http://www.cdc.gov/nceh/vsp/pub/Norovirus/Norovirus.htm

US Public Health Service (2005b), Vessel sanitation programme—operations manual. Atlanta: Centers for Disease Control and Prevention, National Center for Environmental Health.

Widdowson, M.-A., Bulens, S. M., Widdowson, M.-A., Hadley, L., Bresse, J. S., Beard, R. S., et al. (2004), Outbreaks of acute gastroenteritis on cruise ships and on land: Identification of a predominant circulating strain of norovirus—United States, 2002, *Journal of Infectious Diseases*, 190(1), 27–36.

11

Training and Learning on Board

Learning Objectives

By the end of the chapter the reader should be able to:

- Appreciate issues concerning training and learning
- Reflect on learning cultures
- Understand training needs analysis
- Identify a variety of approaches to skills development including training, coaching, and mentoring

This chapter examines the nature of skills development on board. It is argued that training is both essential for operational effectiveness and continuous improvement and development, and, as such, can never be ignored. Training is often managed by the human resource (HR) department but, frequently, it is seen more broadly as an operational function. This implies that training plays a strategic role that is critical for achieving a brand's vision. Training seems to be inseparable from service quality: a company that seeks to achieve excellence in service quality must wholeheartedly embrace training as a key strategic activity.

The provision of training is an investment, and as such there is constant attention on outcomes and benefits, cost effectiveness, and value for money. However, while it is relatively easy to survey passenger feedback about services and to generate scores that reflect satisfaction levels, it can be difficult to measure the impact of training. It is tempting to ascribe changes to single factors, such as training, when the reality may involve a multiplicity of issues. However, training can mean the difference between satisfaction and dissatisfaction, stability and instability, risk and confidence, safety and danger, and profit and loss, and it is a brave or foolish organization that ignores these salient facts. By planning and implementing continuous, effective training, significant improvements may occur, albeit incrementally.

Training differs from learning in that the former is employer directed and the latter is employee motivated. Contemporary practice focuses on the way that organizations can move from a traditional training regime with a "one size fits all" approach to training aimed at creating a learning organization. Learning is the responsibility of the learner, and this latter approach radically changes an organizational culture by recognizing the natural predilection of people to learn and the way that the individual can be encouraged to take responsibility for learning in the workplace to benefit both the individual and the company (Simmonds, 2003).

In identifying key issues relating to training and learning, this chapter will consider training needs aboard megacruise ships and the interface between organizational culture and learning organizations. In addition, a variety of approaches to skills development will be proposed, including training, coaching, and mentoring. Motivation for learning is also considered, and, finally, a case study is presented that considers a cruise company's training provision.

Training and Learning

The task of developing a competent, effective, committed, and passenger-focused workforce on board a cruise ship is demanding. The growth in the industry means that new ships are constantly being constructed, passenger numbers are increasing, passenger's needs are continuously changing, and the need to source the right type of workforce is overwhelming (Wild and Dearing, 2004b). Cruise ships adopt flags or national registrations to create a flexible approach and to limit external control so as to manage costs effectively. This practice is widespread, although by no means universal, and it ensures, among other things, that the labor bill is minimized and that the product is, in turn, available at a price acceptable to the customer. The sourcing of labor from favored countries by using agents was discussed in an earlier chapter. Frequently, the countries selected provide labor that is plentiful and potential employees who are suited to customer service tasks (Dickinson and Vladimir, 1997).

With this pattern of demand for labor, a danger exists that workforce availability will become more problematic, leading to shortages. These shortages mean that HR departments are likely to widen their search for employees to look at emerging sources. Recent patterns have seen employees sourced from the Philippines, India, Mexico, and Eastern Europe. Developing sources are likely to include various South American countries, China, Vietnam, and other Asian countries (Wood, 2000b).

Customer service posts, such as cabin stewards and buffet assistants or waiters, often receive elementary training ashore that is designed to make sure that applicants possess the desired minimum level of dexterity and have good manners with customers, as well as helping employees to adapt to operations when they join the ship. Chefs are frequently sourced from training colleges, although the skill levels achieved may be rudimentary. Selection and recruitment practices result in the contracting of service staff who arrive on board with a need to develop their skills and abilities in the workplace. Standards are attained by a number of interventions, including employing key experienced staff in supervisory posts, constructing manuals that define service and product standards, and training on and off the job. The value of a shore-side training team coupled with peripatetic trainers who travel with ships and offer targeted training can add significant value to this effort (Dickinson and Vladimir, 1997).

Workplace competence is critical for delivering service excellence, but aboard a cruise ship it is only one part of the training equation. Initially, crew must be trained to fit into the shipboard way of life, to orient themselves and find their way around, to settle into the job. Then they are trained to understand and comply with relevant regulations, to develop product knowledge, to deliver expected levels of customer service, to adopt the company and brand culture, to learn how systems work, to know the part to play within a team, and to deal with change. The list is long and challenging. Regulations are vital for safe and secure practice, as are the safety routines associated with emergency drills, and these areas provide a prominent focus for onboard training and an overarching commitment to training across departments.

Learning Cultures

According to Simmonds (2003), trainers play a large part in helping to transform a company from a training-led entity to a learning organization. The subtle advantage in encouraging the individual to take responsibility for personal development, seek opportunities to learn and develop oneself within the organization, and identify one's learning needs creates a paradigm shift that can have far-reaching positive effects. Such an individual is more likely to have high expectations for her or his employment, which should remove barriers for progression or promotion. The learning organization has a specific philosophical approach (see Table 11.1).

This table suggests that a learning organization is one that possesses a vision that is understood and shared by the employees, who are constantly seeking to create improvements, and a strategic aim that seeks to support people as they change and grow. In this organizational climate, employees are valued, supported, rewarded, and appreciated. The organization is sold on the idea of learning and develop-

Table 11.1: **The learning organization, adapted from Honey (2005)**

Beliefs that help define a learning organization

- You cannot make people learn; only make it more likely that they learn.
- Learning and continuous development are too important to leave to chance.
- Complacency is the biggest enemy of continuous improvement/development.
- Most organizations unintentionally reinforce many unwanted behaviors (e.g., deference, blaming, covering up mistakes).
- People learn, not organizations.
- Learning is our core purpose and we sell the results of that learning.
- Learning is the only sustainable competitive advantage.
- Learning has occurred when people can show that they know something they didn't know before and/or can do something they couldn't do before.

Beliefs that hinder when creating a learning organization

- Most of what people learn "just happens" as a natural consequence of doing things and keeping busy.
- The experiences I learn most from are the experiences others will learn most from.
- It is more important to learn from mistakes than to learn from successes.
- Learning from experience mostly happens intuitively (i.e., it isn't necessary to do it deliberately or consciously).
- Developing people is a primary responsibility of any manager.
- The more senior you are, the less you need to learn.
- The learning organization is doomed unless it is done top down.
- Learning is best done at courses, conferences, seminars and workshops.

ment for all who are, in turn, encouraged to introduce ideas. The organization is also open to externally generated ideas.

It may seem almost incongruous that organizations formerly modeled on hierarchical pseudo-militaristic, class-divided structures should adopt a learning organization approach. Yet the evidence suggests that enlightened cruise companies are moving toward this approach in order to derive competitive advantage from their increasingly treasured human resources. For example, Princess Cruises encourages employees to seek career development promotions, and the organization has many examples of individuals who have created opportunities for themselves by taking ownership of their career trajectory and learning opportunities. Within this particular organization, employees can access learning opportunities by signing up for training or using online learning facilities.

The culture of an organization can be described in a number of ways. For example, Evans et al. (2003:81) describe the "cultural web" as relating to:

- Stories that surround the organization that might help to define the people, the successes, or the characteristics.
- Routines and rituals, which for a shipping company might involve the annual Christmas celebration for crew and officers, the crossing the line (equator) ceremony, or the launching of a new vessel. Equally, it can relate to the company's way of achieving service quality through some form of customer service strategy and a related system of rewards.
- Symbols—the logos, the flags, the images relating to sea-going, the use of semiotics or the symbolic representation of signs, and even the shape and color of the ships.
- Structure—the way the ship is organized in teams, departments, and groups of teams, as well as the informal social structures that exist on board and the structure of shoreside organization.
- Control systems, which include budgetary control, quality control, and operational control and can refer to what Hofstede (1986) calls power distance. In this sense, a small power distance environment creates a more liberal, tolerant, and equal society, while an environment with a large power distance has a strict hierarchy and a less liberal regime. Cruise ship societies from different companies can be located along a continuum of power distance.
- Power structures—the location of the head office, the owner of the company, and the prevailing management style that can all feed in to this element of the cultural web.

This cultural web, seen in its entirety, can present a holistic view of an organizational culture. Other views of culture emerge when reflecting back on work undertaken by Handy (1996) who stated that there are four distinct cultural types:

- Power cultures tend to be dominated by an individual or group of people who retain overall control. This type of organization relies on the quality of those key people and whether the organization can respond to change depends on one or a few people.
- Role cultures are frequently hierarchical and give credence to long-standing, well-established procedures and policies. These organizations can appear bureaucratic and can be slow to adopt change.
- Task cultures are often team based and are found when day-to-day routines are replaced by projects or one-off events. The teams are often multiskilled and flexible so as to respond to the demands they face.
- Person cultures are there to support an individual. In this sense such a culture might be a trade union.

Miles and Snow (1978), describe culture for organizations by the way they react strategically. They identify "defenders" as organizations that are generally found in stable, mature markets that frequently occupy a specific niche. These types of businesses defend their territory by targeting costs or making service improvements. Miles and Snow state that such organizations have limited flexibility because of their rather hierarchical style and strict, unyielding control processes. The authors describe "prospectors" as innovators that seek new markets. The organization is aware of the market environment and values flexibility in order to respond to opportunities with appropriate speed. "Analyzers" are careful in approaching the market. This type of culture exists when the business follows the lead of others. Mistakes are avoided by carefully examining data. Finally, "reactors" also take the lead from others but this cultural type is prone to repeating errors. Such a business tends to have poor leadership and unsatisfactory systems.

Training Needs Analysis/Assessment (TNA)

Training needs analysis or assessment can be undertaken for a number of reasons: to remedy a problem, to identify a problem or state of effectiveness, or to take part in an ongoing process of continuous improvement. In the first two cases, the implication that a firm has a training need may suggest that there is an underlying performance problem. In such a situation, TNA can be a form of research undertaken to identify and then address underlying or root causes of problems. TNA can be carried out by an external agency, such as a consultancy, so that the assessment is conducted objectively, without bias, and unfettered by prior or preconceived expectations. It can be done to achieve corporate change, to examine competence levels for potential progression, or to measure the impact of training.

Chiu and Thompson (1999) highlight four distinct methods of data collection in TNA: surveys, individual interviews, focus groups, and on-site observations. The authors note that action research methods, which can capture the true essence of training needs by considering workplace dynamics, are less commonly found despite the potential benefits for an organization that is focused on achieving best levels of performance and meeting strategic objectives. Chiu and Thompson believe that the individual's learning needs are frequently discounted in this type of TNA exercise, but that this is a critical issue because a learning organization cannot afford to lose sight of the individual when they seek to create a balance in meeting both organizational and individual learning needs.

Training needs can also emerge as part of the appraisal process. Appraisals are cyclical (often annual) events. The appraisal process can generate high-quality data about a person in an organization. However, Leat and Lovell (1997) counsel that the process tends to be used differently by different organizations, with the result that in some cases the focus can be on summative performance for the appraisal period, in others the appraisal may reflect on the individual's performance for remuneration or progression, and finally the appraisal may try and equate diagnostic needs against summative judgments.

An effective appraisal will consider the individual, the team, and the organizational needs in a balanced fashion so as to create an action plan. In this sense, a more thorough analysis can create a complete and holistic overview. The organizational analysis will focus on the "organization's goals, skills resources, indices of effectiveness, and the organizational climate" (Leat and Lovell, 1997:149). The task analysis can reflect on the parameters for job roles that are typically identified through a job description and job specification. Finally, the person analysis can address the question about the individual's effectiveness in doing her or his job.

Skills Development

Most jobs include a variety of skills that can be improved by practice, evaluated and developed through comparison with identifiable good practice, or honed to near-perfection through engagement with those who claim "expert" status. A new crew member joining the food and beverage team in a service position is likely to spend a period of time working in crew or officer mess areas before progressing to serve in passenger buffet dining facilities. Thereafter, the individual will proceed to restaurants in a support capacity before taking more responsibility for directing service. Progression is logical, allowing the individual to learn and develop skills through the use of a variety of training interventions and the involvement of key individuals. Learning is cumulative, building on basic skills and routines in order to establish standards that are recognized by the employee's supervisors, managers, and the customers.

On many ships, employees are expected to converse in the predominant language of the cruise brand and the customers. This has as much to do with safety as it does with communication to achieve high levels of customer service. Because the cruise industry is global and attracts a broad range of nationalities and cultures, language can present problems. For example, new hires may have a rudimentary knowledge of English, and their level of language skill may impede their progress from low customer contact areas to high customer contact areas. Language classes may be required on board to help make the employee operationally effective and to ensure that the ship's company complies with safety regulations. Managers should recognize that learning on the job and conversing in a second language can be physically tiring for those who are seeking to develop language skills.

Many skills can be developed on the job with the aid of a set of clear guidelines in the form of a reference manual or manuals that can include pictorial representations of the way a table is laid for meals, the presentation of dishes before they leave the galley, or the standard presentation of toiletries in a guest's bathroom. Lists positioned strategically out of sight of the guests can also act as a useful memory aid. This approach can be supplemented by the use of posters or high-visibility signs to reinforce good practice for issues such as hygiene, food safety, and customer service. On-the-job training can be supported by coaching or timely interventions to make adjustments during service routines. These can be acceptable if the routine is not impeded and service standards do not drop below acceptable levels.

Some skills are better developed away from the customer to allow the individual to practice before using the new skills in operational areas. Off-the-job training can be practiced in training areas or rooms away from the normal work location or within the usual work location but outside normal service times. This form of training can be beneficial when skills require practice and development of technique, when demonstration may be beneficial, when larger groups can benefit from the practice or learning situation, when questions and answers can help confirm details and understanding, or when role-playing can be used to simulate a situation. Invariably, off-the-job training removes risk from the workplace by introducing control features.

Planning a Training Session

There are a number of key steps to follow when preparing to train a small group of staff. It is important to prepare in order to use time effectively and to achieve successful outcomes. The training must be for a purpose that is relevant to meeting set objectives, understandable to the trainee, and manageable within the time and place constraints that exist.

According to Reece and Walker (2000), it is helpful to establish a plan that includes or takes into account:

- Knowledge of trainees—awareness of levels of ability and likely orientation toward the training
- Your objectives, which should be SMART: specific (such as folding serviettes), measurable (to a set standard), achievable (such as giving trainees appropriate basic levels of dexterity), relevant (of direct importance to the job), and timely (to meet a time frame or timetable)
- Timing—to fit in with operational constraints and to help the trainee and trainer concentrate appropriately
- Resources—the conditions and materials that are essential to achieving the objectives (this can include equipment, materials, or teaching aids)
- Training strategy—the approach you will use as a demonstrator or trainer
- Assessment—the approach you will use to check the learning outcome

Reece and Walker (2000) believe that any training session should have as basic components an introduction, a main body, and a conclusion. Each stage plays a part in first setting the scene, clarifying objectives, and stating what will be achieved by the end of the session (introduction); then, delivering the training in a structured, carefully paced and systematic manner while paying attention to critical issues such as hygiene, quality standards, essential knowledge or customer service (main body); and finally, revisiting the objectives, clarifying and stressing what has been covered, assessing learning and checking understanding (conclusion). The training session should encourage attentiveness, so location and conditions related to the place and time where training occurs are important, as well as content and method of delivery. The trainer should raise expectations with the trainees about what they will learn and have high expectations about what they will learn. The trainees should see the benefit that arises from the training both in terms of why they are doing it and how it can help them in their job. The trainees should understand what is expected from them in the session, such as the level of participation, any assessments that are required, and rules that are in agreed to make the session work.

Training is a complex activity that requires that the trainer develop a range of skills so he or she is confident, businesslike, enthusiastic, stimulating, and clear. The relationship with the trainees is usually best when it is mature, professional, and warm rather than aloof and cold. Many skills that are best trained off the job are known as "psychomotor" skills (Reece and Walker, 2000). That is, they require acting or doing. However, it is highly unlikely that a trainee will only learn psychomotor skills in isolation, as invariably the individual will also be expected to comprehend related theory and to understand why they are doing what they are doing.

When planning a psychomotor skill development session, the trainer should be able to analyze the skill and recognize key abilities and describe and demonstrate the skill to clearly portray correct sequencing, coordination and timing. More complex skills may need to be broken down to help understanding. The trainer should then be able to create conditions so the trainee can practice the skill and provide opportunities for feedback to be generated in relation to the learned skill. In some cases, the practice can follow the demonstration step by step before the trainee attempts the task independently. Feedback can be intrinsic (the trainee criticizes her or his own performance), or extrinsic (the trainer or a third party provides the criticism), or both. It is important to provide opportunities for trainees to continue to practice any learned skill so as to increase levels of ability.

Assessment and evaluation can be done through the use of observation and feedback, by asking questions to check understanding, or by using a test (paper, electronic, or practically based). The use of groups and peer assessment can add to this function by providing a support network that creates opportunities for discussion and problem solving. Coaching can be used as a follow-up in the workplace to provide individual support and attention, using expert guidance to focus skills development.

Mentoring

New hires or newly promoted appointees can benefit from the individual attention provided by a mentor in supporting and overseeing the employee's transition from novice to accomplished practitioner. According to Zey (1991), the mentor provides a vital link as a teacher who helps the employee interpret the complex organizational realities, as a counselor to support the employee at times of difficulty, as a sponsor to provide information that can help the employee progress, and as someone who will get involved when necessary to provide protection. The complex environment on board a cruise ship is, it would seem, a fitting place for the use of such a role.

The use of a mentor creates a bridge between the insecurity of a new setting and the social integration and effectiveness of the experienced crew or officer. Problems can arise from within the process if, for example, the wrong type of person is selected to be the mentor or if the mentor or the employee misinterprets the other party. It can be easy to forget that working in any environment creates a new hybrid language that can be unique to the setting. This may have to do with working at sea and the related maritime jargon or the commonality of understanding about routines connected to the job. The mentor should take appropriate care and responsibility for the employee and not use the role as a form of power; equally, the employee should take care to value the role of the mentor and to appreciate the benefits that are accrued from this type of support.

Zey (1991) states that a mentor should be good at his or her job, be supported by the organization, be effective as a teacher and motivator, and be secure in his or her position. Other factors are the ability to empathize with the employee and the accessibility of the mentor. In many respects, these factors are also somewhat academic; because this is a partnership, the individual chemistry that exists is as important as any other element and managers should take care to consider this when aiming for best fit.

Learning and Motivation

It can be argued that as individuals we learn all the time and are constantly adding information and knowledge to our databanks (Gibson, 2003). What we learn will depend on our prior knowledge, beliefs, and the way we interpret our environment or sociocultural setting (Lave and Wenger, 1991). Indeed, in many respects, what may be learned by some individuals could be categorized by an observer as incorrect, inappropriate, or apparently illogical. Learning is a highly personal act, and what we learn is also a matter of our understanding about opportunities in conjunction with the potential to which we think we can aspire (Bloomer and Hodkinson, 2000). In making judgments about learning, the individual considers her or his needs and aspirations so as to rationalize the perceived options for possible action in order to meet a personal set of priorities that exist for the specific context.

The "Circumstantial Curriculum" (see Figure 11.1) encapsulates this theory and is a useful device for understanding learners and their motivation. The theory emphasizes the need to be aware of the way that individuals understand and relate to their setting (Lave, 1988). This can help employers appreciate that different employees may have a different understanding or appreciation of their local environment because of who they are, what they know, how they perceive the realities on board, and their motivations. The model can also help to explain why individuals act as they do in respect to learning opportunities and can suggest ways that trainers can plan learning opportunities to address individual issues. Finally, the theory suggests that because learning is continuous and frequently unplanned, it is important to pay particular attention to the setting on board. This enables the organization to create the best conditions for learning and learners and in turn the most appropriate professional outcomes to suit all parties.

It is useful to consider how this theory can be applied to practice. The following case study presents a training initiative called "White Star Academy" that was introduced by Cunard Lines. When reading this case study, reflect on the learning model and identify how good practice, which is suggested can

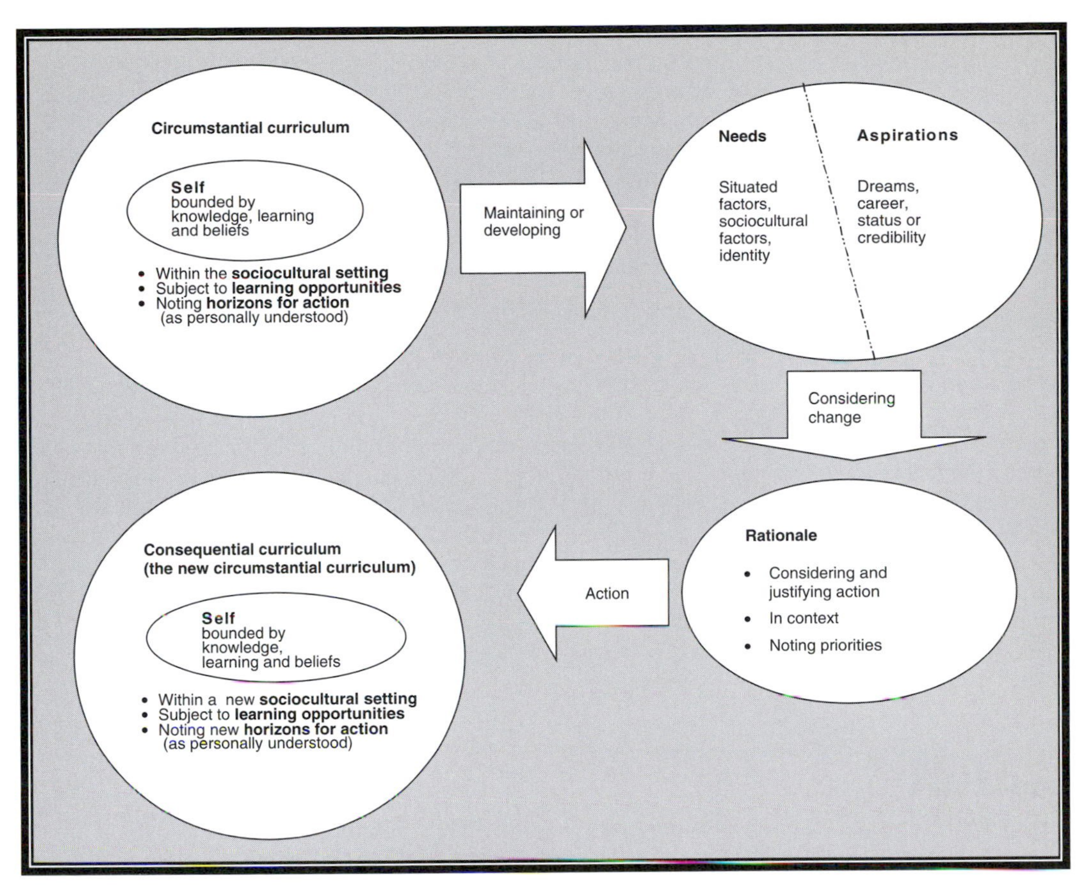

Figure 11.1: The "Circumstantial Curriculum": an integrated theory of learning (Gibson, 2004: 335).

emerge from adopting the understanding implied by the circumstantial curriculum, is evident through the training initiative.

Case Study: White Star Academy (WSA)

Cunard introduced the White Star Academy as a College at Sea in 2000, partially in response to regulations requiring new seagoing staff to receive a comprehensive and compulsory safety induction program. However, in introducing the WSA, Cunard enabled their training team to address skills and knowledge acquisition in a formal and structured manner so as to bolster quality control and provide a more supportive induction for new entrants to the industry. Essentially, the WSA encompasses orientation, induction, diagnostic assessment, the establishment of basic operational skills (including customer service), product knowledge, company information, health, security, safety, and contextual knowledge relating to working at sea.

The academy is structured to allow employees to be led incrementally and logically through a process, developing their learning and preparing them for progression to full operational status. The program is scheduled over four weeks. The various steps allow the employee to start by considering personal appearance, grooming, and the appropriate use of uniforms. This is followed by training relating to the company, working on board, and the specific function the employee will perform. Objectives for each session are clearly stated so the employee knows what is expected to be achieved. The White Star Academy appears to create a powerful focus for training on board that establishes an identity, clarity of purpose, the notion that this is an entity for the protection of high standards, a body of knowledge to differentiate the Cunard brand, and a training regime which, despite a relatively short history, seems to have become an integral part of the Cunard way of doing things.

Summary and Conclusion

Training and development are serious challenges for cruise companies. The pace of growth is such that sourcing of staff becomes a problem. However, honing the ship's complement into an effective operational team is a further consideration. Growth creates opportunities for staff in terms of promotion and personal development, but it can also dilute critical levels of competence. This chapter has examined training issues and reflected on learning issues. The very nature of a ship and the unique society on board is well suited to developing a learning culture. However, in order for such a culture to survive and thrive, the company should consider creating mechanisms to support individuals in their search for learning opportunities. The final chapter presents research findings concerning planning for undergraduate work placements or internships. This is followed by a series of case studies that will help the reader appreciate the range of roles that are relevant in the purser's department.

Glossary

Competence: The quality of being appropriately qualified physically and intellectually.
Diagnostic (as in assessment): To check and identify if there is a learning need.
Education: Encompasses teaching and learning specific skills, and also something less tangible but more profound—the imparting of knowledge, good judgment, and wisdom.
Formative (as in assessment): Assessment that has a primary objective of providing prescriptive feedback.
Learning: The cognitive process of acquiring skill or knowledge.
Paradigm: Refers to a pattern or model; a collection of assumptions, concepts, practices, and values that constitutes a way of viewing reality.
Peripatetic: A person who walks (travels) from place to place in order to do his or her job.
Philosophical: Related to the rational investigation of the principles and truths of being, knowledge, or conduct.
Summative: Focuses on reporting performance or achievement at fixed endpoints.
Training: Activity leading to skilled behavior.

Chapter Review Questions

1. What are the options for training crew on board?
2. Which areas of shipboard life require to be addressed through training because of regulations?
3. What is the difference between a training-oriented organization and a learning organization?
4. How do you design a training session to address practical skills?
5. What are coaching skills?
6. What are mentoring skills?
7. Why is it important to assess learning?
8. How can individual needs be met through a training program?

Additional Sources of Information

www.plymouth.ac.uk: For details of the BSc (Hons) Cruise Operations Management
International Journal of Training and Development
Training magazine
Education & Training Journal
Innovations in Education & Teaching International (*IETI*)
Journal of Education and Work
Journal of Further and Higher Education

References

Bloomer, M., and Hodkinson, P. (2000), Learning careers: Continuity and change in young people's dispositions to learning, *British Educational Research Journal*, 26(5), 583–597.

Chiu, W., and Thompson, D. (1999), Re-thinking training needs analysis, *Personnel Review*, 28(1/2), 77–91.

Dickinson, R., and Vladimir, A. (1997), *Selling the sea*. New York: Wiley.

Evans, N., Campbell, D., and Stonehouse, G. (2003), *Strategic management for travel and tourism*. Oxford: Butterworth Heinemann.

Gibson, P. (2003), *Learning, culture, curriculum and college: A social anthropology*, Unpublished PhD, University of Exeter, Exeter.

Handy, C. B. (1996), *Understanding organisations* (4th ed.). London: Penguin.

Hofstede, G. (1986), Cultural differences in teaching and learning, *International Journal of Intercultural Relations*, 10(3), 301–320.

Honey, P. (2005), *Learning beliefs*. Retrieved May 2005, from http://www.inspiringlearningforall.gov.uk

Lave, J. (1988), *Cognition in practice*. Cambridge: University Press.

Lave, J., and Wenger, E. (1991), *Situated learning: Legitimate peripheral participation*. Cambridge: University Press.

Leat, M. J., and Lovell, M. J. (1997), Training needs analysis: Weaknesses in the conventional approach, *Journal of European Industrial Management*, 21(4/5), 143–154.

Miles, R. E., and Snow, C. C. (1978), *Organizational strategy, structure and process*. New York: McGraw-Hill.

Reece, I., and Walker, S. (2000), *Teaching, training, and learning* (4th ed.). Sunderland: Business Education Publishers.

Simmonds, D. (2003), *Designing and delivering training*. London: CIPD.

Wild, P., and Dearing, J. (2004), Growth culture. *Lloyd's Cruise International*, 17–24.

Wood, R. E. (2000), Caribbean cruise tourism: Globalisation at sea, *Annals of Tourism Research*, 27(2), 345–370.

Zey, M. G. (1991), *The mentor connection*. New Brunswick: Transaction Publishers.

12

Managing Integrated Operations

Learning Objectives

By the end of the chapter the reader should be able to:

- Reflect on research undertaken to establish graduate internships
- Examine case studies or vignettes to reveal insights about working life on board
- Reflect on the administrative role that the purser's office plays on board a contemporary cruise ship
- Consider the complex relationships that together establish integrated operations

So far, this book has presented a series of chapters that have aimed to deconstruct the world of cruising from an operational perspective. In doing so, it has considered the origins of cruising and charted the development of this fast-growing industry. It has reflected on the component parts of cruising to identify what makes a cruise both in terms of the itinerary and the range of services and facilities on board, and it has considered the mechanism for packaging and selling cruises as a product. The book has analyzed the key operational functions for hotel services and the provision of revenue-generating services such as shore excursions. Finally, an overview of security and safety was considered.

This last chapter aims to round off the book by presenting a brief summary of research that was undertaken to plan for graduate internships or work placements on board cruise ships. Thereafter, a series of vignettes or case studies are presented to consider the roles of a range of employees in the purser's department on contemporary cruise ships. In conjunction with the findings from the research project, the vignettes provide the reader with an opportunity to reflect on the day-to-day reality of working for a contemporary cruise brand. The vignettes are written using the voice of the individuals concerned so as to convey authenticity.

Researching Graduate Employment on Cruise Ships

Following the introduction in 2003 of the BSc (Hons) Cruise Operations Management at the University of Plymouth in the United Kingdom, a research project was undertaken to investigate planning for establishing work placements aboard cruise ships. The degree program was designed as a result of collaboration and consultation among the former Institute of Marine Studies (now the School of Shipping and Logistics) at the University, representatives from cruise companies, and members of the Tourism and Hospitality group at the University (Gibson & Nell, 2003). The program was approved and identified as an important qualification for the expanding cruise industry.

This unique undergraduate qualification, with its module mix including cruise operations, management, tourism, maritime studies, and hospitality, was designed to hold relevance for contemporary cruise companies and to prepare future cruise ship hotel services managers for the demands of employment in this setting. An optional one-year industrial placement after stage two was seen as an important component, because students who could secure such a placement would be able to make an informed decision about selecting this type of work as a long-term career. Any student who subsequently rejected the lifestyle relating to this type of work could transfer to another degree qualification in the final year. Thus, the cruise industry would be able to actively engage in offering work placements in the knowledge that the program aided their selection process to identify suitably motivated and prepared candidates. Students who successfully complete a placement and their degree become a secure investment for the cruise company because they are known commodities, will have undergone comprehensive professional development, and are able to make the transition seamlessly from graduate to manager on board.

In conjunction with the introduction of the degree, a research plan was formulated to inform the process of developing placement strategies. The project aimed to create a deep understanding, in relation to life aboard a contemporary cruise ship, for officers and crew in the hotel services department, with a view to developing undergraduate work placements.

Research Planning

In order to achieve the depth of understanding associated to the research aim, the research adopted an interpretive stance with an anthropological focus (Cohen et al., 2000). Qualitative data collection techniques (Denzin and Lincoln, 1998) were applied to the research setting, in order to examine the professional and social domain and to answer two primary research questions: what is it like to live and work on a large contemporary cruise ship, and what are the implications for students studying the recently introduced BSc (Hons) Cruise Operations Management who may be seeking such a work placement?

In May 2004, a series of 24 interviews were conducted over a seven-day period with a range of hotel department employees aboard a cruise ship in the Mediterranean. The sample was carefully selected, in negotiation with senior personnel ashore and on the ship, to give a broadly representative range of hotel "type" and service employees. Thus bars, restaurants, galley and buffet personnel; junior hotel managers; accommodation personnel; shops on board and photography managers; and the cruise director and senior managers, representing a range of nationalities, ranks, and a mix of genders, were interviewed.

The ship that forms the focus for this research was built in 2002. This vessel is generally regarded as typifying a contemporary style of cruise ship (Bjornsen, 2003), a style that is frequently replicated by other cruise brands. The ship carries 2,600 passengers and 1,100 crew. She is operated by a famous international cruise "brand." The ship weighs in at 110,000 GRT and is an example of the largest cruise ships currently in operation.

The interviews were undertaken using a semistructured interview schedule (Bell et al., 1984) with a view to constructing case studies (Bassey, 1999) from the transcripts, so as to gain an insight into this complex world and to reflect on the implications for students on placement. The interviews were carried out with due regard to the setting, conditions, permissions, and ethics (Cohen et al., 2000). The researcher adopted a friendly, encouraging style with open questions and occasional prompts. Interviews were conducted in locations where the interviewees felt comfortable and were scheduled at times that were convenient to the individuals involved. In all cases, confidentiality was offered by withholding the name of the vessel and, in line with respondents' wishes, individuals' names are also disguised.

The transcripts were then transformed into case reports or case studies (Yin, 1994) so as to develop a series of individual narratives that accurately represented the data in an accessible format. The analysis of cases was undertaken using framework analysis within the context of a circumstantial curriculum model (Gibson, 2004). This device (see Chapter 11, Figure 11.1) helps to develop a sensitive interpretation of complex data within a specific setting and to explain motivation, decisionmaking, and action. The elements considered within the framework included the self as a central factor; knowledge, learning and beliefs; the sociocultural setting; options for learning; horizons for action; and needs and aspirations as understood by the individual. Finally, the case study data was scanned to highlight any pertinent issues that affected placement planning.

Results and Findings

A broad picture was produced from the data to throw light on the community on board and the way individuals worked and lived in this unique setting. The findings help to describe the complex "community of practice" (Lave and Wenger, 1991) issues that exist for employees on board this ship. The case subjects were selected in consultation with senior managers from the cruise company to be broadly representative of employees from the hotel department. Because of the link between the students on placement and the types of employment to which they aspired, more time was spent interviewing junior and assistant managers or pursers. Interviewees included departmental managers, APs, JAPs, accommodation supervisors, assistant headwaiters, senior chefs, and assistant waiters.

The case subjects were originally attracted to work on board for a variety of reasons. In most cases, the decision to work on a cruise ship had been partly associated with a desire to travel in connection with the glamor implied by working within a luxury environment. In other cases, the primary attraction had been to work in an environment that provided a good income and a desirable lifestyle. Very few subjects stated simplistic reasons for making the transition from being ashore to working at sea. Responses were generally multidimensional and individually complex. Developing this perspective, it was apparent that most subjects reassessed the working environment as time went by and formulated new reasons for remaining within this type of employment. In some cases, they stated that, while the lure of travel remained, the powerful social context became more important: "My friends are from all over the world and I can visit them in, for example, Canada or Mexico when I go on leave" said a JAP. "I look forward to going home on leave but after a short time I find myself looking forward to coming back," stated an accommodation supervisor. For many subjects, working at sea provides a strong social context that becomes, for whatever reason, an increasingly important element.

The overall picture relating to life on board that emerged was of a community that was considerably more diverse in makeup than had been predicted. There were 54 different nationalities among the crew and a 2:1 male–female ratio. The average age of the crew was in the early 30s. The majority of the subjects were keen to emphasize the community as a model of good practice that existed on board the ship: "It is a lesson to the United Nations that people with different beliefs and different nationalities can live together," said one assistant restaurant manager. The description was apt, as it was easy to observe, in passenger and crew areas, a community in harmony. There were few signs of personal conflict, and those limited occasions where a tension might arise were reported as being infrequent.

It appeared that the community operated at different levels—at any time there were different identifiable communities of practice. The professional level created a milieu within which individuals undertook their role and interacted with supervisors, managers, subordinates, and passengers. The working practices for a waiter were established by attending training sessions, observing colleagues, listening to supervisors, learning from mistakes, and refining routines through practice. A JAP adopted similar approaches when establishing working practices, although in addition, this subject had access to online learning materials, had been provided with customized training, and was periodically updated by line managers in relation to product knowledge and corporate practices. Senior managers adopted their routines, relating working practices to cumulative learning, corporate missives, standard operational procedures, and the need to respond to demands from head office. There was a sense that the working environment had changed because of electronic communication (email, and the intranet and Internet), which meant the professional community at sea was less isolated and subject to more external observation.

The subjects regarded the shared environment on board in much the same way that residents view a village or town. There was a feeling of ownership presented by some as "this is a special ship to work on" or of loyalty and pride expressed by others as "this ship is the best ship to be on; better than any other." The time spent on board was divided between "on duty" and "off duty." Privileges afforded to some personnel, such as officers and managers, created an opportunity for them to socialize in uniform when they were not undertaking routine tasks. In other words, they were still on duty but in a more relaxed mode. While crew, staff, and officers or managers were not allowed in passenger areas out of uniform, in some cases, as long as the person wore a badge, this was considered sufficient. Working hours were long, reflecting the nature of the operation and the need to maintain continuity of services.

It appeared that the professional community of practice was directly affected by the way that managers managed. In a sense, this was a reflection of the management style adopted by individual managers. But at a more contemplative level, the "mood" on the vessel indicated that a patterned approach to the way that the personnel on board communicated with each other, while conducting professional duties, was, in some way, derivative of the lead provided by senior managers. Managers and supervisors appeared to conform to their interpretation of the "norm" in deciding how to treat people and how to communicate with their teams. There appeared to be evidence of two-way learning, in that managers actively learned about the people they worked with, and the staff actively learned about their managers and their environment. In many respects, this finding adds another dimension to a study undertaken by Testa (2002), which implied that nationality and cultural background affect leadership for the cruise industry. The case studies suggested that individuals were sensitive to the complex cultural differences that existed and that, at different times, decisions were made to subsume strong cultural beliefs or expectations to maintain the harmonious working balance.

This patterned approach carried forward to the social communities of practice within which members formed social alliances and networks. The networks were frequently, but not exclusively, formed according to hierarchical, national, or cultural similarities. Occasional examples of what was described earlier as "infrequent tensions" were sometimes glimpsed through the cases, when a member of a social community encroached on what was seen to be a territorial boundary. For example, a table in the crew mess was habitually claimed by a group of Mexican cabin stewards. It was noticeable that early contacts, formed when crew members first joined the ship, were important in helping to orient the individual and in allowing that person to settle in.

There was a sense that the community was self-aware; individuals stated their understanding about the complex makeup of this group of staff on board, and that this awareness was important in creating harmony. The hierarchy on board was understood and yet, because of the American model of cruising where the nautical is subsumed by the vacational, the pseudomilitaristic version of naval officers at sea was less prevalent. Rules and regulations play a significant part in setting societal parameters, but the subjects had clearly absorbed the meaning of regulatory life on board. The ship operated effectively because there was a collective will to make sure the social and professional context met each individual's need. The subjects expressed an understanding that they were being paid to do a job and, if they were not successful, the job would disappear. It was in each person's best interest to maintain a balance. Individuals who were unable to comply appeared to be in the minority, and were identified and repatriated very quickly. This ship appeared to be a perfect example of a self-regulating society.

The numerical bias toward males on board seemed to create certain issues worth noting. To begin with, a female joining the ship found that she soon became the focus of male attention. This focus was an exaggerated version of what might happen in a typical shoreside situation. Females in this position reported that the attention could, at various times, be flattering, irritating, annoying, and exasperating. Subjects described the techniques they would use to ward off unwelcome approaches and that eventually, after a period of time, the attention became less of an issue.

There were many examples of couples working on board. Sometimes they were married or in close partnerships. The company appeared to be flexible in meeting the needs of individuals to work and live with a partner. This included making arrangements for cabin accommodation. Social conditions were described as being very good. An assistant waitress said, "I'm not allowed in passenger areas when I am not working. It sounds like I am something less than the passengers, but I don't care, the facilities are good. There is a pool, a Jacuzzi, a gym, and the crew bar. I like being with my friends. They are a mixture of Poles, Mexicans, and Romanians."

Subjects described the speed with which they found themselves settling into the community on board. After a period of leave, rejoining a vessel or proceeding to a new vessel created few concerns because the contractual and employment patterns meant they inevitably met people with whom they had worked before. Friendships were quickly re-established. In many respects, the communities on cruise ships were described in a manner that made them sound like university campus communities. The community members work in an environment where levels of interaction and communication are high. This means that individuals cannot become isolated and problems cannot be ignored. The success of the social community appeared to be important to enabling success for the professional community.

It is apparent that the social communities on board different cruise ships may well display similarities, but each ship is regarded as being unique. Indeed, as one senior assistant purser notes, "the ships are different, the types of people, the passengers, the size, the itinerary—people want to be there for these reasons." Her interpretation about the reasons that ships are popular (from her point of view and that of her colleagues) stresses that the uniqueness of the individual setting arises from the complexity of the variables relating to each setting. The dynamics on board are affected by changing rotations (as contracts end and crew members go on leave); the demographics of the passengers; the cruise itineraries; the way that individuals regard the physical environment of the ship; and the manner in which senior managers manage the ship.

Implications

Interns or placement students are in an interesting position when joining a cruise ship community. In some respects, the community will be distinctly unsettling because it is likely to be a self-replicating, self-balancing, and self-regulating environment quite unlike that experienced onshore. That said, with careful planning, the individual can make a rapid transition to fit in and to feel comfortable.

It appears to be important that the placement student is provided with clear induction information relating to rules and regulations so that the working environment is clearly understood. Also, to facilitate the learning progression from new inductee to a settled crew member, techniques should be considered to help create networks. These can include appointing appropriate mentors, holding group inductions with shared team orientation sessions, and developing team-building activities. An orientation cruise that was offered in May 2004 to University of Plymouth students planning to join a cruise company in July 2005 was an important element for the first group of placement students to achieve these ends.

The cultural context may be different from what the individual has experienced, so some form of cultural induction should be considered to introduce factors for recognition. These can include awareness related to cultural differences, suggestions about settling in, what to expect, and directions about who to contact to get help or advice.

The study suggests that there are considerable strengths within this type of community and that the natural human predilection to learn is the critical element for achieving a successful integration. The organization is most likely to succeed in this task by considering how it can create the best conditions to help the learner to learn. There is a countersuggestion connected to the findings, which anticipates that the working dynamics may not be as successful on some ships as on others. There are obvious implications in studying such communities to identify critical factors so as to make recommendations for good practice.

The Purser's Office and Integrated Practice

The purser's office on a cruise ship is the administrative hub for the vessel. The office is usually fronted by a reception desk, which is the focal point for passengers to interface with the cruise company on board. The type of contact is unpredictable because, while a passenger may discuss food and wine with a restaurant manager or accommodation matters with an accommodation steward, the passenger sees this desk area as a shipboard version of an Internet Search engine such as Google. Any question may be asked and the answer is expected promptly and accurately. The nature of the contact can range from the simple to the complex, but, throughout, the service must remain consistent and the level of professionalism high.

Donna, Senior Assistant Purser (SAP) Front Desk

This first vignette considers a manager who oversees the front desk. In 1996, Donna McBride joined as a junior assistant purser for Princess Cruises with the full intention of staying for 6 months. She had been working previously with travel agents and hotels but was attracted to the glamour of travel. Now she is a senior member of the administration team responsible for the front line work at the purser's office, managing a team of assistant pursers (APs) and junior assistant pursers (JAPs). The work is constant and demanding but she also describes it as incredibly rewarding. She reports to the staff first purser (Administration) and then to the passenger services director. She controls a team that operates the purser's office and ensures that the front desk is managed effectively.

Donna works closely with her APs to provide 24-hour service. One AP acts as the night manager, another is front desk supervisor, and the "Pratika" is an AP who facilitates administrative matters such as port clearance and dealing with port officials. The front desk also acts for the medical center to cover the incoming phone calls for the ship's doctor. Her duties also mean she is in regular contact with the accounts manager and many other managers on board. Her team includes 12 JAPs who are on rotation to ensure the front desk is staffed. They also cover additional duties as may be required of them, including assisting the night manager, administering the art auction (Art Track), lost property, assisting with work placement, captain's circle, and the casino. There are usually four or five JAPs on the front desk at any time depending on the itinerary and the volume of passengers seeking assistance. The office operates from 0800 to 1900 with a lunch break. The assistant night manager covers the office 1900 through to 2300 when the night manager is on duty through to 0800. JAPs tend to be on duty for 11 hours each day and 7 or 8 hours when in port. Turnaround days tend to be the busiest days.

Much of her time is divided between coordinating the front office duties, ensuring that passengers' problems are solved, and training her staff to develop customer service skills, IT skills, and product knowledge. Training can be one-on-one or small groups. Training in product knowledge covers every aspect of front office operations, the Princess product, and systems training. Depending on whether it is the JAP's first contract, there may also be a need for general refresher training, updating, or familiarization. Additionally, Princess operates "Princess U." with the U. standing for university, which is an online training program to help shipboard personnel improve their service skills, develop team-building strategies, and deal with problem solving.

The front desk is a magnet for passengers who have something to say. On a ship with over 2,000 passengers there are many different types of people, some of whom may appear illogical or who are asking the impossible. Yet her team must be consistent, friendly, passenger focused, and dedicated to satisfying the passenger. Frequently problems are presented that are easily solved. Sometimes the problem is more difficult. A focus file is a computer-generated report that tracks a recorded problem from beginning to end. Different itineraries can create different problems. In the Caribbean, passengers can often be accompanied by large quantities of baggage and have problems if an item goes missing. In Europe, US passengers have flown long distances and there have been occasions when their baggage does not arrive with them at the ship. Focus files are sent ashore to the passenger relations department to create a record in case follow-up is required. The ship's staff is not empowered to make refunds: that is the domain of the corporate office.

Donna recalls that people can get very upset if they think their luggage has gone missing. She describes people wandering around in dressing gowns, swearing or crying. People can often be at their worst in these situations, but the JAPs helping the passengers must always do their very best to help.

Eventually, passengers calm down and may even apologize if they have behaved badly. It can be that her team does its best but the customer does not always see it that way. If luggage is not located in time for sailing, the JAP working with the AP or SAP will follow the complaint through until it is resolved. Then arrangements will be made to get the luggage forwarded to the next port and the passenger will be helped with temporary solutions, in terms of the provision of clothes to wear (including formal wear) and express laundry service. The intention is to make the passenger as comfortable as possible.

Passengers can complete comment forms that go to a drop box to be collated by the Captain's secretary. In turn, these are copied to the PSD and then sent to the corporate office. Every element of the cruise is reported on. The aim on board is to keep customer satisfaction scores as high as possible, although these are not always reflective of real issues and can be a representation of the types of passengers, the weather, or specific issues that were out of the control of cruise personnel.

The diversity of customer complaints is worth considering. The front office team is often faced with making judgments about genuine complaints in sometimes rather unusual circumstances. For example, the night manager tends to have interesting situations to deal with. People seem to develop different personalities at night, possibly as a result of drinking too much alcohol. On one occasion the night manager was summoned to attend when a female passenger had collapsed outside a lift and her husband had refused to have anything more to do with her. Passengers have even been known to threaten suicide. Whatever the situation, it has to be dealt with carefully, compassionately, and effectively. The company has a strict policy that any passenger threatening or being violent would be sent off at the next port.

While passenger services are administered at the front desk, the Pratika administers passenger records, deals with cabins that are empty because passengers have not arrived (or have cancelled), and ensures that details are available for port authorities. The night manager prepares guest folios, ensuring that they are up to date and that records are maintained in case of a query.

Donna believes that cruise ships all feel different in terms of the working and living atmosphere and the culture on board. This may be because of the crew, the passengers, the size of the ship, or the itinerary. She believes that people on board have their own individual reasons for working on the ship but, irrespective of this, everyone will probably feel differently because of the ship they are on. The ship's dynamics are complex. The company tried to put together a team of like-minded people to set the ship up and, subsequently, they use teams to replicate the setup. As contracts expire and people change, the ship may change. The managers are equally important. They set a benchmark regarding the tone of the leadership style. On this ship there is a relaxed style, everyone appears to be happy in his or her work and life on board, and people are trusted to get on with their jobs. There appears to be an opposite potential in that those who are too strict may create a negative atmosphere that can affect passengers.

The CRUISE credo makes Princess different to all other cruise brands. While it has changed over time, it has also become more and more important. Princess has developed competitions with prizes such as employee of the month. The cruise committee review blue (generated by crew) or green (generated by passengers) feedback forms and award the accolade complete with financial reward. Personnel who complete Princess U. courses successfully can also be rewarded financially.

In Donna's view, the cultural aspect of working in the front office does not really affect life on board. The people in her office are multinational and represent a broad range of cultures. The only practical issue that can arise may relate to language and understanding. This is because some strong accents can affect understanding and some JAPs may also speak too quickly, thus inhibiting comprehension. In addition, passengers and crew may incorrectly interpret rudeness from the way someone says something. When the ship is cruising in the Mediterranean there are significant advantages in having an office team that is multilingual.

Donna's story includes a range of insights concerning life and work on board. Readers are prompted to reflect on the continuity of coverage, the need to create a depth of coverage for specific days and times, the importance of customer feedback forms, and the subtle implications of the cultural milieu on board.

Vince, Senior Assistant Purser (SAP) Accounts

This next vignette considers the accounting function on board. Vince is a hotel management graduate from Italy. He started his employment career with the company as a junior assistant purser five years ago and has found opportunities for promotion to be very good. He attributes this to the exceptional growth that the industry has experienced, coupled with the trend to construct large ships. He describes operations on board these vessels as dynamic and he identifies the challenge faced by contemporary cruise companies to source and retain the right staff.

His position as an onboard accounts manager means he works closely with a number of senior personnel, including the SAP front office, the crew SAP, the staff first purser (Administration), and the department managers for all revenue areas. On smaller ships, the roles of accounts manager and SAP front office are merged, and located in a large office dominated by a large and impressive safe that contains currencies that may be required on the voyage. The ship is virtually cash-free because the majority of passengers pay their folios or accounts by credit card, so the amount of money on board is considerably lower than would otherwise be the case. Cash, in the form of bank notes, is used to pay the crew (US dollars) and to stock currency transaction machines (actual currencies will depend on the itinerary). He has at his disposal a note counter and a coin counter, which he shares with casino staff.

His routines include managing the shipboard accounts, checking that guest folios are updated, posting or recording figures relating to all sales on board (passengers and crew), reconciling the various financial records to balance the cruise accounts, allocating floats for reception staff and any other cashiers who require a float, allocating cash for payment of crew wages, and allocating gratuities. He works with a shore-based accountant who has the responsibility of monitoring and checking the financial health of the vessel, cruise by cruise, and follows standard procedures using standardized documentation. He is also responsible for the various currencies that are carried on board and for preparing and managing the automatic change machines.

The busiest day tends to be when the ship returns to the home port. Vince gets up at 0330 for a start time of 0430. He has to prepare all figures to "close the cruise." This term is used to identify that point when all transactions and records relating to the cruise are completed and the records are sent to the head office. That morning, final bills go to passengers (passengers can check folios halfway through so there is no shock). At 0600, passengers start disembarkation. Vince notes that the office can get hectic for the next hour or more as passengers visit the purser's desk to query their folio. Passengers leave by 1000, and this creates a two-hour window when Vince completes the cruise closure routines and prepares to open the records for the next cruise by 1200.

While relatively rare, mistakes can happen. A sale may have been miscoded or allocated to the wrong passenger account. Computer errors can occur, but there is a technician on board who can rectify that, if necessary. Despite the large number of passengers on board, there may be 15–20 mistakes. Any error changes the cruise account and these must be handled with care. Refunds on errors are approved by senior staff and corroborated by the departmental manager who is responsible for the transaction. Vince takes the view that if a mistake cannot be found, the issue can become embarrassingly public, so it is for him a matter of pride to be accurate and timely.

Vince aims to prepare a document box that contains all accounting paperwork. This is sent to head office. During the day, cash may be offloaded using a contracted arrangement with a security firm. Money may also be ordered and delivered using the same method. The itinerary will determine the types of currencies that are to be carried on board. The company provides a currency exchange service using automated machines. Some revenue may be generated in selling and buying back notes but the provision is primarily intended to be a service. During this stage of the day, Vince will balance his float. His experience has shown him that discrepancies can arise when closing the cruise because a number of individuals can input data into the accounting software. This can sometimes lead to a data input error and as a result, Vince has to trace the source of any imbalance. Fraud is not a serious issue. Credit cards are checked, cashiers are trained to identify forged notes, and the system of using credit cards raises on-board security. Junior assistant pursers (JAPs) are meant to have cash-handling training prior to appointment, but some training is done on board, and all receive training prior to joining.

Usually 8–10 appointees spend two weeks aboard a ship going through theory, policy, and product training prior to joining their first ship. Vince states that it is important for JAPs to possess as much knowledge as possible about the itinerary; practical matters, such as the currency ashore; and the products on board.

Vince does not feel isolated in his job, as there is a lot of interaction. He emails head office frequently and gives revenue managers the figures to show patterns and trends. For every cruise there is a revenue meeting to check targets relating to what is known as Passenger Berthing Daily (PBD) spending. Vince finishes his day in the afternoon prior to departure at around 1500 or 1600. The new cruise is ready, accounting systems are prepared, and floats are allocated to all cashiers. It is a long day, but he is satisfied to have sent his documentation to the head office on time and in the correct state of accuracy.

Vince's role is central within the administrative system. His remit means he deals with a wide variety of different managers and colleagues on board and ashore. There are a number of officers on a cruise ship that act in this way, providing essential services to support operational effectiveness.

Tanya, Senior Assistant Purser (SAP) Crew

This next vignette describes the role of the crew senior assistant purser, who undertakes a key role in the human resource management function. Tanya studied tourism marketing at a hotel school. She enjoyed her course but felt that it was too theoretical and didn't include enough practice or training from industry. She came to this conclusion having worked in hotels, restaurants, and in administration. Working at sea is very different from the equivalent hotel job. For example, at sea, the reservation process is divorced from the hotel administration function, while, in a hotel, reservations are a significant element.

Tanya has worked for this company for five years. She has been a JAP and has experienced lots of different jobs: the crew office, working on the Purser's desk, shore excursions, and assistant night manager. Then she was appointed AP for two contracts before being promoted to SAP. In the crew office, she is responsible for tasks relating to the 1,100 personnel on board. These tasks include wages, welfare, staffing, induction, and administration.

She is supported in her job by an AP and a JAP. The crew office is located centrally in the crew area along a corridor colloquially known as the M1, which is the main service artery through the ship. Crew come to this office to find out information, to get help with personal matters, to collect items or wages, to arrange details for returning home, and to communicate with the ship's personnel department. Most crew are paid in cash, and some have money wired home, which Tanya and her staff deal with. They can also supply debit cards for use by personnel in the crew bar or messes. The crew office role calls for the team to work closely together and to trust each other. This is the first place the crew will come to talk to the "company," which calls for the team to deal with each person with sensitivity, respect, and tact. A recent situation occurred when a crew member's mother had died and the company arranged to repatriate him. The office had to organize, in consultation with other colleagues, to obtain the necessary visa and to arrange a flight.

All heads of department request new staff via the crew office. Each request is passed on, in turn, to the head office. New personnel receive an induction pack from the crew office, are given induction information about life on board, and are then provided a series of induction training sessions that cover the use of watertight doors, conduct and regulations, and finally the use of fire extinguishers and lifeboat drills. The deck department undertakes all safety inductions. Every two weeks zone commanders or people in charge of stretcher parties train their teams.

The crew office team maintain a set of accounts; they act as a bank for the crew, dealing with all financial matters. In addition they sell phone cards that can be used on phones that are available for crew use only. There is a lot of paperwork connected to the job. Passports are held when the crew join and returned when they leave. A security system is employed which is used to register when a crew member (or for that matter, a passenger) leaves and then returns to the ship. This system generates a pass with a bar code that identifies the holder. A photograph is generated that can be checked by security personnel at the gangway. If the crew member does not return, Tanya has to ensure that the passport is provided for the shoreside agent to hand over to the crew. The multinational crew is interesting because of the way they view on board life, how they adapt to become culturally sensitive and integrated, and how people can change as a result of working on board. Dealing with problems can be difficult. Crew members come to the crew SAP expecting Tanya and her team to solve everything. Unfortunately, as Tanya states, "Life isn't always like that." Tanya believes that the crew respect the officer's uniform but also expect the responsibility or authority of the uniform to solve their problems.

Tanya holds down a very responsible and extremely important job. She likes sea days because for her and her team there is less to go wrong. Every time the ship is in port there is a risk that a member of the ship's company will miss sailing, and because each port can bring a new country, it can bring a new set of rules that can add complexity to the situation. Consider what you would do if there were 20 minutes to sailing and five crew members have not returned on board.

David, Assistant Purser: Pratika

This next vignette examines the role of the Pratika. This function is inextricably linked with the task of entering and leaving countries, crossing borders, and dealing with the transportation of people and goods. The growth of cruising has increased the volume of passengers and crew who are traveling by cruise ship and has created the need to focus on the processes that occur to gain clearance for the ship's passengers and crew to leave a ship when the vessel arrives in port.

David deals with all local authorities to get the ship cleared by both customs and immigration officials. Some itineraries are easier than others because of the rules and regulations they have in place and the way these rules and regulations can be interpreted. David believes that the United States is strict in terms of their rules, which he believes are interpreted vigorously by officials. In Italy, ships may find that officials seek different types of information not only if the ship visits different ports but also even when the ship returns to a previously visited port. In Europe, a number of ports and countries are covered by the Schengen agreement (see Chapter 5).

The Pratika has an overview covering everything that goes on and off the ship. David coordinates with the ship's agent, customs officers, and immigration officials. The work requires that preparation be done before the ship arrives in port and that lists of passengers and crew are sent to the ship's agent for processing the day before arrival. This inevitably means that the home port day, or turnaround day, is a particularly challenging time, especially if the ship arrives in a port the day after. Visas can cause problems because passengers can sometimes ether fail to get a visa or get the wrong visa. There are many variables, and, if officials say the visa is invalid, the passenger cannot disembark.

Pratika is an Italian term, which in the cruising world means "paperwork," and David notes that this is a big part of the job. David receives passenger lists from the head office and he also has to deal with updates to these lists; special requests relating to reservations, upgrades, or changes to cabins approved by the head office; and special information about specific groups of passengers. His role includes facilitating clearance by providing information to the relevant authorities and enabling embarkation by monitoring and assisting at check-in, dealing with cabin requests, and assigning emergency cards for non-revenue passengers. He then supports disembarkation by assigning luggage tags to priority early flights, manages groups, and edits and distributes the internal routine and disembarkation schedule. He also looks after arrangements for transit passengers.

The role of the Pratika is a big responsibility. David says he has learned a lot in his time on the job. He has learned to organize and, sometimes, to improvise. He has learned how to nurture partnerships with port agents, port officials, and customs and immigration. The hours of work can be long and the role can be stressful and unpredictable—things can be going well then something emerges at the last moment. His day starts on a port day after the ship docks and port officials come aboard where the crew purser meets them. The Pratika lays out the necessary paperwork, including passports, in a lounge area. When the port official is satisfied he or she announces that the ship is clear.

The Pratika will supply passenger lists, lists about embarking and disembarking passengers, crew lists, bonded stores list, notices of any goods or supplies coming aboard, and a list from the environmental officer regarding any items being discharged or offloaded. The Pratika will make use of the shoreside agent to act as an intermediary in the process. The agent can be a cultural interpreter to facilitate the process. David is assisted in his job by a JAP who takes on the data input and administrative duties that help him to divide his time appropriately to ensure his targets are achieved.

Recently, a passenger had a severe heart attack. She had to be transported to a shoreside hospital immediately. David tried to arrange for an ambulance but, when it did not arrive, he contacted an emergency helicopter service. He then had to undertake delicate negotiations with the passenger's traveling companion to find out what she wished to do. This resulted in the companion leaving the ship and moving into a hotel close to the hospital. Working with the agent and colleagues, David tried to make this traumatic event as stress free as possible. Carrying large numbers of passengers is likely to mean that problems can emerge that must be dealt with instantly and appropriately so as to satisfy all parties.

The Pratika is a member of the purser's department who acts on information sent to the ship from the head office, then processes that information to ensure the cruise operates to plan. The term *Pratika* is not common to all cruise companies, and the reader should be aware that other job titles could be used.

Cherie, Assistant Purser Hotel Services

The person described in this final vignette also takes on this type of activity, where the employee is an interpreter and facilitator.

Cherie looks after special services on board. This covers a variety of tasks, from family and friends who wish to send flowers or gifts to passengers, to arranging weddings on board or shoreside, and to coordinating special celebration packages. She works almost independently, reporting to the staff first purser and the passenger services director (PSD). Requests for special services arrive from the head office with the Pratika and they are then sent to her. She creates a schedule to ensure the various requests are dealt with, any package that is ordered is organized, and all departments are fully instructed in terms of requirements.

The role is relatively new and has been introduced in response to growing popularity for special services. The post is not gender specific; the previous AP who took on the role was a male officer. The job is rewarding and, as with many of the JAP/AP/SAP jobs, very demanding. When the schedule becomes complex she can call on the PSD's secretary to assist her.

Cherie has three events this particular day. She is at work by just before 0800 and immediately starts double-checking arrangements. Any event is a special event for Cherie and she wants to make sure it is perfect for those involved. There are two weddings and a renewal of vows scheduled. The timing is quite tight, but, by carefully coordinating reception venues and with the support of her colleagues, everything goes according to plan. Weddings are often customized, with the bride and groom requesting packages to suit their personal preferences. The bride can start her day by having her hair styled and being pampered in the beauty salon. Cherie makes sure the florist sends the appropriate flowers to the salon, checks that the bar where the welcome reception is held is prepared, and that the wedding chapel is ready.

When she is in her office she checks that certificates are prepared, confirms the arrangements with the captain, double-checks that the photographer is prepared, and starts to orchestrate events. Each wedding and event must flow seamlessly, with the aim of ensuring that the focus is placed where it should be. The AP hotel services becomes a wedding planner and spends a lot of time leading the process, guiding all parties, and ensuring the event stays on schedule. There are many details to be noted, and for that reason the role suits the individual who is meticulous and who has particularly strong interpersonal skills.

More and more large cruise ships possess a wedding chapel that can be used for a variety of formal events, complete with live or recorded music and decorated appropriately. An added service is provided by the broadcast of pictures from the wedding ceremony to the company web page. This enables friends and family to be a virtual audience to the celebration. Cherie highlights the importance of the Captain's role in the proceedings. The personal touch and the Captain's care, concern, and professionalism help to put a special touch on the occasion. Cherie really gets into the day and admits to brushing the odd tear away. She stresses that she really enjoys what she does, attributing this to the nature of what is for all a happy and positive experience.

After the ceremony, which can be religious or secular, traditional or nontraditional, the wedding group will have photographs taken around the ship; this may include the tradition of cutting the cake. An onboard celebration can continue through the cruise if requested, with all sorts of opportunities for customization and the creation of special moments. At the end of the cruise, Cherie will send the paperwork to the head office for onward dispatch to the country where the ship is registered in order to formalize any wedding event. A card is sent to the happy couple to congratulate them and to pinpoint the location at sea where the event occurred.

On a port day, Cherie catches up with her other duties, which may include ensuring the delivery of gift vouchers and dealing with any ongoing requests for special items. In addition to her high-profile work, she also acts on the PSD's behalf to help induct new crew members for hotel services. This includes allocating emergency cards and ensuring that the crew members know their emergency duties, orientation of the area, and ensuring they sign the Captain's standing orders. Finally, she also acts for the PSD to ensure that all policies and procedures are updated and that these are then circulated and signed as read by the appropriate personnel.

Integrated Operations: Conclusion

The management and administration of hotel services on contemporary cruise ships are not easily understood from the passenger's point of view. The ship's personnel engage in a range of activities and tasks that are designed to ensure that operations are continuous, that high levels of customer service are constant, that the passenger is the focus, and that all services are available. In order to achieve these objectives, the individuals working on board must function as a team. Moreover, emerging developments highlight the need for the onboard team and the shoreside team to share common aims and objectives so as to become congruous.

The research findings and the vignettes in this chapter stress the notion of integration in practice. Working at sea is not for those who cherish isolation. It is possible to steal moments of peace and solitude, but the environment is more suited to those who value human interaction and who empathize with others.

Glossary

Anthropological: A term related to the study of humankind and human culture.
Case study: A method for learning about a complex instance, based on a comprehensive understanding of that instance, obtained by extensive description and analysis of the instance, taken as a whole and in its context.
Ethics: The philosophical study of moral values and rules.
Folio: Another term for a passenger's account.
Google: A popular search engine.
Interpretive: Inductive reasoning.
Milieu: An inclusive term for a setting or environment.
Qualitative (as in research): Observation of research phenomena in situ; that is, within their naturally occurring context or contexts.

Chapter Review Questions

1. What are the elements that combine to define the cruise ship as a community of practice?
2. What are the lessons that emerge from the research project to inform graduate interns?
3. Identify, for each role described in the individual vignettes, the points of contact that require the post holder to liaise with other crew members.
4. What are the challenges for those working on the purser's desk?
5. What currencies are required for itineraries including ports in Mexico, Italy, Greece, Turkey, Alaska, Australia, and China?

Additional Reading and Sources of Further Information

Bassey, M. (1999), *Case study research in educational settings*. Buckingham: Open University Press.

Bell, J., Bush, T., Fox, A., Goodey, J., and Goulding, S. (1984), *Conducting small-scale investigations in educational management.* Milton Keynes: P-C-P Open University.

Bjornsen, P. (2003), The growth of the market and global competition in the cruise industry. Paper presented at the Cruise and Ferry Conference, Earls Court, London.

Bow, S. (2002), *Working on cruise ships*. Oxford: Vacation Work Publishing.

Cohen, L., Manion, L., and Morrison, K. (2000), *Research methods in education* (5th ed.). London: RoutledgeFalmer.

Denzin, N. K., & Lincoln, Y. S. (Eds.) (1998), *Collecting and interpreting qualitative materials*. Thousand Oaks, Ca.: Sage Publications.

Gibson, P. (2004), Life and learning in further education: constructing the circumstantial curriculum, *Journal of Further and Higher Education*, 28, 333–346.

Gibson, P., and Nell, J. (2003), Professional development and hotel services on cruise ships. Paper presented at the Cruise and Ferry Conference 2003, Earls Court London.

Gibson, P. (2005), Communities of practice: Employment on cruise ships. Paper presented at the CHME Research Conference, Bournemouth.

Lave, J., & Wenger, E. (1991), *Situated learning: Legitimate peripheral participation*. Cambridge: University Press.

Testa, M. R. (2002), Leadership dyads in the cruise industry: The impact of cultural congruency, *International Journal of Hospitality Management*, 21(4), 425–441.

Yin, R. K. (1994), *Case study research*. Thousand Oaks, Ca.: Sage Publications.

Index

N

O

P

Q

R

S

T

U

V

W

Y